U0180747

剑桥植物学术语图典

（英汉对照）

〔英〕迈克尔·希基（Michael Hickey）
〔英〕克里夫·金（Clive King） 著

于胜祥　侯元同　郭成勇　侯春丽　译

The Cambridge Illustrated Glossary
of Botanical Terms

北京大学出版社
PEKING UNIVERSITY PRESS

著作权合同登记号 图字：01-2013-3704

图书在版编目(CIP)数据

剑桥植物学术语图典：英汉对照/（英）迈克尔·希基（Michael Hickey），（英）克里夫·金（Clive King）著；于胜祥等译. —北京：北京大学出版社，2021.10
ISBN 978-7-301-23932-2

Ⅰ.①剑… Ⅱ.①迈… ②克… ③于… Ⅲ.①植物－名词术语－图集 Ⅳ.①Q949-61

中国版本图书馆CIP数据核字（2014）第022566号

书　　　名	剑桥植物学术语图典（英汉对照） JIANQIAO ZHIWUXUE SHUYU TUDIAN（YINGHAN DUIZHAO）
著作责任者	〔英〕迈克尔·希基（Michael Hickey）　〔英〕克里夫·金（Clive King）　著 于胜祥　侯元同　郭成勇　侯春丽　等　译
责 任 编 辑	巩佳佳
标 准 书 号	ISBN 978-7-301-23932-2
出 版 发 行	北京大学出版社
地　　　址	北京市海淀区成府路205号　100871
网　　　址	http://www.pup.cn　新浪微博：@北京大学出版社
电 子 信 箱	zyjy@pup.cn
电　　　话	邮购部 010-62752015　发行部 010-62750672　编辑部 010-62704142
印 刷 者	北京中科印刷有限公司
经 销 者	新华书店
	787毫米×1092毫米　16开本　18.75印张　453千字
	2021年10月第1版　2021年10月第1次印刷
定　　　价	98.00元（精装）

CAMBRIDGE UNIVERSITY PRESS
Cambridge, New York, Melbourne, Madrid, Cape Town, Singapore, São Paulo,
Delhi, Dubai, Tokyo, Mexico City

Cambridge University Press
The Edinburgh Building, Cambridge, CB2 8RU, UK

Published in the United States of America by Cambridge University Press, New York

www.cambridge.org
Information on this title: www.cambridge.org/9780521794015

First published 2000
11th printing 2013

Printed and bound in the United Kingdom by the MPG Books Group

A catalogue record for this publication is available from the British Library

ISBN 978-0-521-79401-5 Paperback

译者序

　　植物分类学文献卷帙浩繁，其中的植物形态术语更是复杂难辨。近年来对植物学感兴趣的人越来越多，这些复杂难辨的植物形态术语给大家带来了极大的困扰。

　　在中国植物分类学发展的初期，植物形态术语多以汇编形式出现在一些拉丁文植物学词汇中，后来英文植物形态术语著作相继问世，但大部分只配有极少量的插图，不方便学习和使用。所以图文并茂的植物形态术语辞书越来越受到植物学爱好者的欢迎和喜爱。

　　本书原作者迈克尔·希基和克里夫·金在剑桥大学植物园主要从事植物学及植物科学绘图方面的科普工作。本书包括了常用的2400余条植物形态术语，又配有精美的植物形态解剖图，对这些形态术语进行解释说明，既体现了科学性，又体现了艺术性，故本书更适合植物分类学爱好者以及初学者作为学习入门的工具书。

　　在本书的翻译过程中，译者参考了之前出版的绝大部分有关或涉及植物形态术语方面的著作，包括一些重要的英语辞书，力求在术语的翻译上与前人保持一致。但对原书中存在问题的个别术语，译者在翻译时尽可能做了修改。例如，原书词汇部分将"bisulcate"(两沟的)误写为"disulcate"，将"hirsutulous"（稍具硬毛的)误写为"hirsutullous"，在"asymmetric"（不对称的）的解释中将"than"误写为"then"，在"bisexual"（两性的）的解释中将"pistils"误写为"carpels"，在"cupuliferous"（具壳斗的）的解释中将"nut"误写为"seed"，在"dichogamous"（雌雄蕊异熟的）的解释中将"pistils"误写为"stigmas"，等等。原书图解部分，Conifer1中将"seed scale或cone scale［种（果）鳞］"误认为是"Ovuliferous scale"(珠鳞)，（因为图示中种子已成熟，珠鳞已发育成熟变成种鳞或果鳞），将"Microsporangium (with pollen grains)"误认为是"Microsporangium (with stamens)"，小孢子叶即雄蕊，小孢子囊在小孢子叶的下面，小孢囊不可能具有雄蕊，应该具花粉粒；Conifer Allies 3中将直立买麻藤胚珠顶部的"Micropylar tube"误认为是"Stigma"；Fruit 7中将已发育成果实（苹果）的"carpopodium"（果柄）误认为是"pedicel" (花柄)；Structure of Grasses 3中将"Long-awned lemma"（具长芒的外稃）与"Lower glume"（下颖片）的位置标反；Hairs and Scales 3中误将"Stellate"（星状的）图片标注为"Pannose"（具毡毛的）；Stem Shapes,

Leaf Fall, Apical Meristem, Phototropism中误将"Vascular bundle scar"（维管束痕）标注为"Vascular bundle"（维管束），等等。

　　本书的翻译出版得到了北京大学出版社的大力支持，感谢陈斌惠先生和巩佳佳女士所付出的辛劳。

　　本书译者水平有限，所翻译的内容难免会有不足之处，敬请读者批评指正！

译　者
2021年10月

FOREWORD

Most glossaries are found on a few pages at the end of text-books, floras, monographs, and other botanical works, and these are often only partially illustrated, and sometimes not at all. We have tried to fill this gap by producing an independent and well illustrated glossary that includes all the terms most commonly used in describing vascular plants, as well as some that are found in more specialised works.

In addition to morphological terms, referring to parts of plants visible with the naked eye or with a hand lens of $\times 10$ magnification, we have included some elementary histological, cytological, and genetical terms found in the more general books on botany.

Chemical terms, however, have been largely excluded, as it was felt that these were outside the scope of this book.

The definition of each term has been kept brief, but this has been supplemented wherever possible by a line drawing.

The illustrations are for the most part original, and have been drawn from living material. Many of these have a naturalistic rendering, though some, by necessity, are diagrammatic. In a few cases, we have had to make adaptations of existing drawings, either because suitable specimens could not be obtained, or because a particular illustration could not be improved upon.

Although the terms are listed in the traditional alphabetical order, the illustrations have been grouped according to their content in order to make the arrangement more "user-friendly". It is hoped that its large format will make the book clear and easy to use, and that readers at all levels of understanding, both amateur and professional, will find it helpful in their chosen area of study, especially plant sciences, horticulture, field studies and botanical illustration.

前　言

　　绝大多数的术语表只是出现在教科书、植物志、专著或其他植物学著作的最后几页，它们通常仅具有部分图解，有时完全没有。本书试图制作一个独立而且配有精美图解的术语表来改善这种状况，这些术语包括在维管植物描述中最为常用的术语，以及在更为专业的著作中见到的术语。

　　除了形态学术语——通过肉眼或手持10倍放大镜可以看到的植物体部分的术语，本书还包括了在更为综合的植物学著作中出现的有关初级组织学、细胞学和遗传学的术语。

　　然而，化学术语很大程度上是被排除在外的，因为我们认为这些内容已超出本书的范围。

　　本书对每个术语的定义力争简洁，同时尽可能用线描图来补充说明。

　　本书的大部分插图是原创，而且是依据活材料绘制的，虽然有些是概略性的，但大多数都是自然状态的。在少数情况下，我们不得不对现有插图做出修改。这样做一方面是因为无法获得合适的标本，另一方面是因为无法对特定的插图进行改进。

　　尽管本书的术语部分仍以传统的字母顺序列出，但为了更方便读者，我们已经按照它们的内容进行了组合。我们非常希望这种大开本能使本书更加清晰易用，同时希望本书对所有认识层次的读者（不管是业余的，还是专业的），在他们各自的研究领域，特别是植物学、园艺学、野外研究和植物绘图等方面都有所裨益。

PREFACE

The collaboration between Michael Hickey and Clive King, authors of this new Glossary, has become a very familiar facet of the life of the Cambridge University Botanic Garden, and the small part I have been able to play over the years in furthering this productive partnership has given me continuing pleasure.

This entirely new book represents a logical further step in the concern of the authors to supply the increasing educated public interested in the naming and classification of vascular plants with the tools of their trade. Technical terms, derived almost exclusively from the classical languages of Latin and Greek, are indispensable if we wish to proceed beyond a very superficial knowledge of the great variety of plants in gardens or in nature, and I feel very confident that this carefully constructed illustrated Glossary will meet all our needs-including my own, as I continue, in old age, to enjoy my expert hobby as much as I ever did.

I commend the book unreservedly to you, the next generation, for whom it is designed.

S. M. Walters,
Formerly Director,
University Botanic Garden,
Cambridge

ACKNOWLEDGEMENTS

We would like to express our appreciation and thanks to Dr. S. M. Walters for his encouragement and helpful comments, and for consenting to write the Foreword to this book. We would also like to thank Professor John Parker, Director of the University Botanic Garden, Cambridge, for enabling us to use the facilities within the Botanic Garden and Cory Library, as well as for spending his valuable time looking through the manuscript. We are also grateful for the advice generously given by Dr. James Cullen, Director of the Stanley Smith Horticultural Trust, and by Caroline Poole, teacher of biology at Cheltenham Ladies' College, and for the help provided by the staff of Cheltenham and Gloucester College of Higher Education.

In particular we would like to thank Robert King for his expert use of computer equipment which resulted in the production of the camera-ready copy, and also Diane Hudman, who had the complex job of labelling all the illustrations.

Finally, we are indebted to Dr. Maria Murphy of Cambridge University Press for her much appreciated advice and assistance with the layout of this book.

序　言

　　这本新图典的作者迈克尔·希基和克利夫·金之间的密切合作已成为剑桥大学植物园生活中一个非常熟悉的榜样。多年来我所能做的，能够深化这种富有成效伙伴关系所尽的绵薄之力给我带来了源源不断的快乐。

　　这本全新的书代表了一种必然的进步，越来越多受过教育的公众对维管植物的命名和分类感兴趣，作者为他们提供了这本行业工具书。如果我们希望超越基于在植物园和自然界生长的植物之间巨大变异的那些非常肤浅的知识，那么从传统拉丁语和希腊语演变而来的那些专门术语将是不可或缺的。同时我非常确信这本细心编纂的配有图解的术语词典将满足我们所有的需要。像我一样，虽然年迈，但仍与以前一样在享受我的专业爱好。

　　我毫无保留地向大家推荐这本书，因为它是专门为你们设计的。

斯图亚特·马克斯·沃尔特
剑桥大学植物园原主任

致　谢

　　对于斯图亚特·马克斯·沃尔特博士的鼓励和富有成效的建议，并愿意为本书写序，我们向他表达诚挚的谢意和感激。我们还感谢剑桥大学植物园主任约翰·帕克教授，他为我们充分利用植物园和核心图书馆提供了便利条件，同时花费精力为我们审阅书稿。我们也感谢史丹利·斯密斯园艺公司总经理詹姆斯·卡伦博士和切尔滕纳姆女子学院生物学教师卡洛琳·普尔的热情建议，以及切尔滕纳姆与格洛斯特高等教育学院的同事们提供的帮助。

　　我们尤其感谢熟练操作计算机设备制作影像材料的罗伯特·金，还有承担标记所有图版复杂工作的戴安娜·胡德曼。

　　最后，我们感激剑桥大学出版社的玛丽亚·墨菲博士为该书的设计提供了很多建议和帮助。

The Cambridge Illustrated Glossary of Botanical Terms
剑桥植物学术语图典

This comprehensive and beautifully illustrated glossary comprises over 2400 terms commonly used to describe vascular plants. The majority are structural terms referring to parts of plants visible with the naked eye or with a ×10 hand lens, but some elementary microscopical and physiological terms are also included, as appropriate. Each term is defined accurately and concisely, and whenever possible, cross referenced to clearly labelled line drawings made mainly from living material. The illustrations are presented together in a section comprising 127 large format pages, within which they are grouped according to specific features, such as leaf shape or flower structure, so allowing comparison of different forms at a glance. In addition to supporting the definitions, the illustrations therefore also provide a unique compilation of information that can be referred to independently of the definitions. This makes the glossary a particularly versatile reference work for all those needing a guide to botanical terminology, equally useful to beginners as well as to those more advanced in their botanical knowledge.

MICHAEL HICKEY trained in botany and horticulture, including a period at the University Botanic Garden, Cambridge, before embarking on a teaching career that has spanned over 40 years, ranging from teaching biology to school children through to instructing adults in basic botany and the art

本书内容全面，图解精美，收载了2400余条用来描述维管植物的常用术语。主体部分是关于植物体各器官的结构术语，这些器官都是人们通过肉眼或手持10倍放大镜能够看到的。但在适当情况下，本书也包括了一些基本的显微术语和生理学术语。对每一术语，本书都给出了准确而简明扼要的解释，还尽可能地交叉引用了清楚标注过且基于活材料绘制的线描图。本书将所有图解集中放在一起，共142页，同时，根据具体特征，如叶形或花结构等，又将相关图解进行了一定组合，以便不同形态术语之间的快速比较。为了进一步支持定义内容，图解部分还提供了一个独特的涉及术语定义独立性的信息部分。对于所有需要植物学术语的学者来讲，本书不失为一部特别丰富多彩的工具书，同时，本书对初学者和高级学者在植物学知识方向同样有用。

迈克尔·希基（Michael Hickey），受训于植物学和园艺学专业，教学生涯达40余年，从主讲学校儿童的生物学至教授成年人的基础植物学和植物学绘图技术，在此之前还在剑桥大学植物园工作过一段时间。因其钢笔墨水画曾荣获皇家园艺学会林德利金银奖章，他的作品曾在

of botanical illustration. Awarded the Royal Horticultural Society's Silver Gilt and Silver Lindley Medals for his pen and ink drawings. he has exhibited in the UK and the USA and has contributed illustrations to numerous publications. He is author of *Plant Names: A Guide for Botanical Artists* (1993), *Drawing Plants in Pen and Ink* (1994), and *Botany for Beginners* (1999). With Clive King, he has co-authored *100 Families of Flowering Plants* (1981, 1988) and *Common Families of Flowering Plants* (1997).

CLIVE KING spent over thirty years at the University Botanic Garden, Cambridge as Assistant Taxonomist and Librarian. During that time he was involved in the identification of plants belonging to a wide range of families, the botanical instruction of the annual intake of students, and the day-to-day running of the specialist library housed there. He has translated *Collins Guide to Tropical Plants* (1983) and *Collins Photo Guide to the Wild Flowers of the Mediterranean* (1990), and has co-translated *Collins Photographic Key to the Trees of Britain and Northern Europe* (1988), all from the original editions in German. With Michael Hickey he has co-authored *100 Families of Flowering Plants* (1981, 1988) and *Common Families of Flowering Plants* (1997).

英国和美国巡展，他还曾为许多著作绘制过插图。他是《植物名称：植物学艺术家指南》（1993）、《钢笔墨水绘制植物》（1994）和《启蒙植物学》（1999）的作者。他还与克里夫·金合作出版了《有花植物100科》（1981, 1988）和《有花植物常见科》（1997）。

克里夫·金（Clive King），作为助理分类学家和图书管理员在剑桥大学植物园工作30余年。期间他专心致力于许多科植物的鉴定，负责学生每年的植物学教学和专家图书馆的日常事务。他从德文原版翻译了《科林斯热带植物指南》（1983）和《科林斯地中海野生花卉摄影技术》（1990），与他人合译了《科林斯不列颠和北欧树木摄影要点》（1988）。与迈克尔·希基合作出版了《有花植物100科》（1981, 1988）和《有花植物常见科》（1997）。

Contents
目录

Notes to Readers
读者须知

1. Symbols, prefixes, and suffixes precede the glossary, but abbreviations are found under their appropriate letter.

2. In some cases, an alternative term (in round brackets) occurs after the term listed. This may either be followed by an explanation of the term or it may indicate where the definition may be found.

3. The number [in square brackets] after the definition of a term, or the relevant part of it, shows the page on which there is an illustration of the term concerned.

4. Where a term is used in more than one context, or the authors feel that additional drawings would be helpful, several page references are given.

5. As far as possible illustrations referring to a particular part of a plant have been placed on adjacent pages, beginning with roots and other underground parts and moving upwards through stems and leaves to flowers and fruits.

6. Certain pages have been devoted to plant families or classes in which vegetative or floral structure requires the use of special terms.

7. An arrow has been used to indicate a particular feature in illustrations where possible confusion might arise.

8. Practical studies will often need the use of a hand lens with ×10 magnification. In a few cases, e.g. when observing cell tissue, spores, or pollen grains, access to a microscope will be necessary (see illustrations below).

1. 符号、前缀和后缀列在词汇表之前，但缩写词则列在相应词条中。

2. 在某些情况下，互用术语（圆括号内）在对应术语后面列出。这样，后面既可以跟术语解释，也可以表示定义所在位置。

3. 术语解释或其相关部分后面的数字[方括号内]表示与术语相关的图解所在的页码。

4. 一条术语用在一处以上的上下文中，或作者觉得另外的图解将更有帮助，则列出数个页码以供参考。

5. 关于植物体特殊部分的图解尽可能排列在相近页面上，从根和其他地下部分开始，向上经茎和叶，到达花和果实。

6. 某些页面应用于营养或花结构需要应用特殊术语的植物科或纲。

7. 箭头用来表示图上可能引起混淆的一个特殊特征。

8. 实际研究通常需要用10倍手持放大镜。在极少情况下，如当观察细胞组织、孢子或花粉时，还须使用显微镜（见下图）。

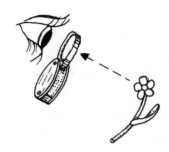

Using a hand lens

Hold the lens close to the eye, and in good light move the specimen towards the lens

手持放大镜的用法

手持放大镜接近眼睛，在光线充足的条件下，移动标本至放大镜

Compound optical microscope

Approximate magnification range: ×40 to ×60

Note: Advanced models magnify up to ×1000+

复式光学显微镜

近似放大范围：×40—×60

注意：更先进的可达到×1000+

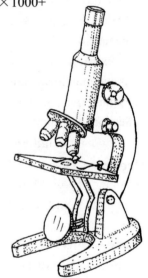

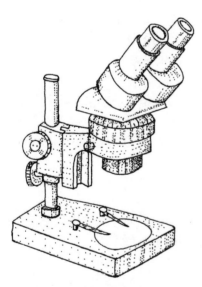

Zoom stereo microscope

Approximate magnification range: ×7 to ×40

Note: Cheaper models without a zoom facility are also available

变焦体视显微镜

近似放大范围：×7—×40

注意：没有变焦设备的便宜式样也可用

Symbols

+ (before a botanical plant name) Graft hybrid

+ (between numbers)Indicates that the plant parts concerned are in two groups

- (between numbers) Indicates the range of variation

± More or less

> More than

< Less than

≥ At least, not less than

≤ At most, not more than

() (enclosing a number) Connate

⌒ (above two letters) Adnate

∞ Indefinite number

♂ Male

♀ Female

☿ Bisexual, hermaphrodite

⊕ Actinomorphic, regular

·ı· Zygomorphic, irregular

§ Section, or other divisions of a genus

符号

+ （位于一个植物名称之前）嫁接杂种

+ （位于数字之间）表示被关注的植物部分分为两组

- （位于数字之间）表示变异的范围

± 或多或少

> 多于

< 少于

≥ 至少，不少于

≤ 至多，不多于

() （包括一个数字）合生的

⌒ （两个字母之上）贴生的

∞ 不定数

♂ 雄性的

♀ 雌性的

☿ 两性的

⊕ 辐射对称的，整齐的

·ı· 两侧对称的，不整齐的

§ 组或一个属下的其他划分

Prefixes (General)

a-	Not, without
acro-	Uppermost, farthest
amphi-	On both sides, around
andro-	Male
aniso-	Unequal
anti-	Against, opposite to
apo-	Separate
bathy-	Lowest
circum-	Around
chloro-	Green
chori-	Separate
chromo-	Colour(ed)
crassi-	Thick, dense
dialy-	Separate
e-	Without
eco-	Ecological
endo-	Within
epi-	Upon, above, over

前缀（一般的）

a-	不，没有
acro-	最上端的，最远端的
amphi-	在两面，周围
andro-	雄性的
aniso-	不等的
anti-	逆的，对立于……
apo-	离生的
bathy-	最下面的
circum-	周围、附近
chloro-	绿色的
chori-	离生的
chromo-	有色的
crassi-	厚的、密的
dialy-	离生的
e-	没有
eco-	生态的
endo-	里面的、内面的
epi-	上面的

ex-	Without	ex-	没有
exo-	Outside	exo-	外面的
extra-	Outside	extra-	外面的
gamo-	United	gamo-	合生的
gymno-	Naked, exposed	gymno-	裸露的
gyno-	Female	gyno-	雌性的
haplo-	Single	haplo-	单倍的、单一的
hetero-	Different	hetero-	不同的
holo-	Whole	holo-	整体的、全部的
homo-, homoio-	Same, alike	homo-, homoio-	相同的、类似的
hypo-	Beneath, below	hypo-	下面的
infra-	Below	infra-	下面的
inter-	Between, among	inter-	之间、中间
intra-	Within, inside	intra-	里面、内面
iso-	Equal, same	iso-	相等的、相同的
macro-	Large	macro-	大的
medi-	Middle	medi-	中等的
mega-	Large	mega-	大的
meso-	Middle	meso-	中等的
micro-	Small, very small	micro-	小的、非常小的
multi-	Many	multi-	多数的
n-, notho-	Hybrid	n-, notho-	杂种的
ob-	Inverse(ly)	ob-	倒
oligo-	Few	oligo-	少数的
ortho-	Straight	ortho-	直立的
palaeo-, paleo-	Ancient	palaeo-, paleo-	古老的、古代的
pauci-	Few	pauci-	少数的
peri-	Around, enclosing	peri-	周围的
pleio-	More	pleio-	多数的
pluri-	Several	pluri-	一些
poly-	Many	poly-	多数的
post-	After	post-	后面的
pro-	Before, in front of, earlier	pro-	前的、早的
proto-	First	proto-	首先的、第一的
pseudo-	False	pseudo-	假的
schizo-	Split	schizo-	分离的
sub-	Beneath, less than, approximately	sub-	在下面、在底下，少于，接近
super-, supra-	Above	super-, supra-	上面、上部
sym-, syn-	With, together	sym-, syn-	和、合、一起
tenui-	Thin	tenui-	细的、薄的

Prefixes （Numerical） 前缀 （数量的）

	LATIN 拉丁文	GREEK 希腊文	CHINESE中文
½	semi-	hemi-	½
1	uni-	mono-	1
2	bi-	di-	2
3	tri-	tri-	3
4	quadri-	tetra-	4
5	quinque-	penta-	5
6	sex-	hexa-	6
7	septem-	hepta-	7
8	octo-	octo-	8
9	novem-	ennea-	9
10	decem-	deca-	10
11	undecim-	endeca-	11
12	duodecim-	dodeca-	12
20	viginti-	icosa-	20
100	centi-	hecato-	100
1000	milli-	chilio-	1000

Suffixes （details of taxonomic ranks are given in the Glossary） 后缀 （分类等级的详细情况在术语汇编中给出）

-aceae	Family, a taxonomic rank	-aceae	科，分类等级
-ales	Order, a taxonomic rank	-ales	目，分类等级
-ara	Added to a personal name to indicate an intergeneric orchid hybrid	-ara	加在人名之后表示兰花的属间杂种
-eae	Tribe, a taxonomic rank	-eae	族，分类等级
-ferous	Bearing, producing	-ferous	具有……，产生……
-fid (leaves)	Divided up to half-way towards the rachis or base	-fid (leaves)	（叶）中裂，裂至羽轴或基部的一半
-foliolate (leaves)	Consisting of a certain number of leaflets	-foliolate (leaves)	（叶）包括一定数目的小叶
-idae	Subclass, a taxonomic rank	-idae	亚纲，分类等级
-inae	Subtribe, a taxonomic rank	-inae	亚族，分类等级
-ineae	Suborder, a taxonomic rank	-ineae	亚目，分类等级
-merous (flowers)	Having the parts of the flower of a particular number, or a multiple of that number	-merous (flowers)	（花）基数，具有一个特殊数字的花的部分，或那个数字的倍数

-oid	Resembling	-oid	像……，类似于……
-oideae	Subfamily, a taxonomic rank	-oideae	亚科，分类等级
-opsida	Class, a taxonomic rank	-opsida	纲，分类等级
-partite (leaves)	Divided from half to two-thirds of the way towards the rachis or base	-partite (leaves)	（叶）深裂，裂至羽轴或基部的一半至三分之二
-phyta	Division, a taxonomic rank	-phyta	门，分类等级
-phytina	Subdivision, a taxonomic rank	-phytina	亚门，分类等级
-ploid (genetics)	The number of sets of chromosomes in each cell	-ploid (genetics)	（遗传学）每个细胞内染色体的套数
-sect (leaves)	Divided almost to the rachis or base	-sect (leaves)	（叶）全裂，几乎裂至羽轴或基部

Measurements

度量衡

c. (circa)	About, approximately	c. (circa)	大约，接近
m	Metre(s)	m	米
cm	Centimetre(s)	cm	厘米
mm	Millimetre(s)	mm	毫米

Taxonomic Ranks

分类等级

The following list of taxonomic ranks is arranged in descending order. Details of the ranks are given in the Glossary.
Principal ranks are shown in bold type, e.g. **kingdom**, secondary ranks in roman type, e.g. tribe, and additional ranks (all with the prefix 'sub-') are in italics.

下面分类等级列表是按照递减顺序排列的。等级的详细情况在词汇中给出。
主要等级以粗体显示，例如界，次一级等级以罗马体显示，如族，额外等级（所有具有前缀"sub-"）以斜体显示。

kingdom 界
subkingdom *亚界*
  division 门
  subdivision *亚门*
    class 纲
    subclass *亚纲*
      order 目
      suborder *亚目*
        family 科
        subfamily *亚科*
          tribe 族

subtribe	亚族
genus	**属**
subgenus	亚属
section	组
subsection	亚组
series	系
subseries	亚系
species	**种**
subspecies	亚种
variety	变种
subvariety	亚变种
form	变型
subform	亚变型

Glossary
词汇

Latin plurals: The following examples show how the plural is formed for most of the Latin terms listed below.

Less common Latin plurals and Greek plurals, where required, are given after the terms concerned.

SINGULAR	PLURAL
locul-us	**locul-i**
pinn-a	**pinn-ae**
haustori-um	**haustori-a**

A (in a floral formula) Androecium, e.g. A10 indicates an androecium composed of 10 stamens.

abaxial The side of an organ away from the axis, dorsal. **[188]**

aberrant Not typical, differing from the normal form.

abortion Non-formation or incompletion of a part. **[258]**

abortive Imperfectly developed.

abscission The shedding of parts of a plant, e.g. leaves, flowers, etc. by means of an abscission layer, either naturally from old age or prematurely from stress. **[173]**

abscission layer A layer of cells that develops across the base of a petiole or pedicel and then weakens, causing the leaf or flower to fall off. **[173]**

acanthophyll A spine, often large, derived from a leaflet. **[248]**

acaulescent Without a stem, or apparently so. **[165, 247]**

accrescent Increasing in size with age, as

拉丁词复数：下面的例子说明以下列出的大多数拉丁词复数是怎样形成的。

在需要时，一些不常见的拉丁词复数和希腊词复数，列在了相关词的后面。

	单数	复数
室	locul-us	locul-i
羽片	pinn-a	pinn-ae
吸器	haustori-um	haustori-a

雄蕊群（在花程式中）Androecium的缩写，例如A10表示雄蕊群由10枚雄蕊组成。

远轴的 远离轴的器官面，背面的。[188]

异常的 非典型的，不同于正常型的。

败育 未形成或不完全的一部分。[258]

败育的 未完全发育的。

脱落，脱离 植物体各部分，如叶、花等，由于自然衰老或压力作用过早地通过离层脱落。[173]

离层 生长于叶柄或花梗基部的一层细胞，当其变弱时引起叶或花脱落。[173]

叶刺 一种通常大、由小叶变态而来的刺。[248]

无茎的 没有茎或貌似没有茎。[165, 247]

渐增的，花后膨大的 随着年龄增长而体积

the calyx of some plants after flowering, e.g. *Physalis alkekengi* (Chinese Lantern). **[210]**

逐渐变大，像花期后某些植物的花萼，如酸浆（中国灯笼）。[210]

accumbent Of cotyledons, having the edges adjacent to the radicle. **[160]**

依伏的，紧贴的　子叶的边缘倚着胚根。[160]

achene A small, dry, one-seeded, indehiscent fruit, strictly of one carpel, as in the genera *Ranunculus* and Rosa. **[257]**

瘦果　小、干燥、具一枚种子、不开裂的果实，严格地一心皮，像毛茛属和蔷薇属的果实。[257]

achlamydeous Without a perianth, as the flowers of *Salix* (willow). **[211]**

无花被的　没有花被，像柳属植物（柳树）的花。[211]

acicle A stiff bristle or slender prickle, sometimes with a gland at its apex. **[180]**

针，刺　硬刺或细刺，有时其顶端具腺体。[180]

acicular Needle-shaped. **[190]**

针状的　针形的。[190]

acid soils Soils with a pH value of 6.5 or below.

酸性土壤　pH值在6.5或以下的土壤。

acidophile A plant that thrives in an acid soil.

嗜酸植物（嗜酸生物）　一类在酸性土壤上苗壮生长的植物。

acidophilic, acidophilous Adapted to or thriving in an acid soil.

嗜酸的，喜酸的　适应于酸性土壤的或在酸性土壤上苗壮生长的。

acorn The fruit of the genus *Quercus* (oak) in the Fagaceae. **[184]**

橡果　壳斗科栎属植物（橡树）的果实。[184]

acrogen A flowerless plant, e.g. a fern, in which growth occurs only at the apex of the stem.

顶生植物　隐花植物，如蕨类植物，仅在茎顶端生长。

acrogenous Growing only at the apex of the stem.

顶生的　仅从茎顶端进行生长。

acropetal Produced, developing, or opening in succession from base to apex.

向顶的　从基部到顶端依次产生、发展或开放的。

acrophyll One of the mature fronds of a climbing fern that occur in the upper part of the plant. **[277]**

顶生叶　攀缘蕨类植物的茎轴上部产生的成熟叶状体之一。[277]

acroscopic On the side towards the apex. **[278]**

向上的　在向着顶端的一边。[278]

acrospire The first sprout of a germinating seed.

初生叶　萌发种子的第一个芽。

acrostichoid Resembling *Acrostichum*, one of several genera of ferns in which the sporangia are distributed over the lower surface of the fertile lamina. **[279]**

卤蕨型　类似于卤蕨属的，该属是孢子囊分布于能育羽片背面的数个蕨类植物属之一。[279]

acrotonic A type of branching in which the shoots nearest the apex of the stem show the greatest development.

顶端优势　一种分枝的类型，最接近茎顶端的芽表现出最大的发育优势。

actinomorphic (regular, radially symmetric) Divisible through the centre of the flower in several or many longitudinal planes, the halves of the flower being mirror images in every case. **[209]**

actinostele A type of protostele, in which the xylem forms a star-shaped structure with phloem between its rays. **[280]**

aculeate Bearing prickles. **[180]**

acuminate Narrowing gradually to a point. **[194, 195]**

acute Sharply pointed. **[194, 195]**

acyclic Arranged spirally rather than in whorls.

adaxial The side of an organ towards the axis, ventral. **[188]**

adherent In close contact with a different part, but not fused with it.

adnate United with a different part, as stipules to a petiole **[178]**, or a bract to a peduncle. **[184]**

adpressed (see **appressed**).

adventitious bud A bud that arises from any part of a plant other than the axil of a leaf.

adventitious root A root that arises from any part of a plant other than the primary root system. **[145, 147, 150, 152, 158, 168]**

adventive Growing spontaneously in a particular region but not native there.

aerenchyma Tissue with well-developed air spaces between the cells, characteristic of the roots and stems of water plants. **[246]**

aerial root An adventitious root that does not grow down into the soil, e.g. the roots of epiphytic orchids that absorb water from the surrounding air. **[146]**

aerotaxis Movement in response to the source of oxygen.

aerotropic Turning towards or away from the source of oxygen.

aerotropism The growth movement of a plant

辐射对称的（整齐、辐射对称） 以几个或多个径向面经花的中央分开，任一情况下花的每一半都是镜像。 [209]

星状中柱 一种原生中柱类型，其中木质部形成星状结构，星射线之间具有韧皮部。[280]

具皮刺的 具有皮刺的。[180]

渐尖的 逐渐变窄成一点。[194，195]

急尖的 急剧变尖。[194，195]

非轮生的 螺旋状而非轮状排列的。

近轴的 靠向轴的器官面，腹面。[188]

附着的 与不同部分紧密接触，但并没有融合。

贴生的 以不同部分合生，像托叶合生到叶柄 [178]，或苞片合生到花序梗。[184]

紧贴的（见"紧贴的"）。

不定芽 一类发生于叶腋以外的植物其他部分的芽。

不定根 一类发生于主根系以外的植物其他部分的根。[145, 147, 150, 152, 158, 168]

外来的，非本土的 自然生长在一特定地区，但不是原生的。

通气组织 一类水生植物根和茎内特有的细胞间具有发育良好气空的组织。[246]

气生根 一类没有向下生长到土壤内的不定根，如从周围空气吸收水分的附生兰的根。[146]

趋氧性 向有氧气的地方运动。

避氧性 朝远离氧气的地方运动。

向氧性，向气性 向有氧气的地方生长

in response to oxygen.

aestival Occurring in early summer.

aestivation The arrangement of the calyx or corolla in a flower bud. (see illustrations of **vernation** for terms used)

aff. (affinis) Having affinity with, near to. Usually precedes the name of a species to indicate a plant not conforming exactly with the description of that species but clearly related to it.

agamospermy (seed apomixis) A form of apomixis in which seed is set, but without sexual fusion. Offspring produced in this way have the genetic constitution of the parent plant. Genera in which this process occurs include *Taraxacum* (dandelion), *Hieracium* (hawkweed), and *Rubus* (blackberry, etc.).

agg. (aggregatum) Aggregate, added to the name of a species to signify the inclusion of other taxa in a closely related group.

aggregate fruit A fruit formed by the joining of several carpels that were separate in the flower, as in the genus *Rubus* (Rosaceae). **[257]**

aggregate species (collective species) A group of two or more closely related species which for convenience have been given a shared name, e.g. *Rubus fruticosus* (blackberry).

ala A wing; one of the two lateral petals in the flowers of plants of subfamily Papilionoideae in the Leguminosae. **[230]**

alar (see **axillary**).

alar flower A flower borne in the fork between the two branches of a dichasium, e.g. as in some genera of the Caryophyllaceae. **[204]**

alate (winged) Having a wing or wings. **[159, 173, 177, 258]**

albumen Nutritive material stored within the seed.

albuminous Possessing albumen.

运动。

夏季的 生长于初夏的。

花被卷叠式 花萼或花冠在花芽中的排列方式（详见"幼叶卷叠式"词的图解）。

近缘的 与……相类似，近似于……通常用在某植物名称前面，表示该植物并不完全符合那种植物的描述，但与之明显相关。

无配子种子生殖（种子无融合生殖） 一种无融合生殖的形式，形成种子但无性融合。以这种方式产生的后代含有亲本的遗传物质。属于该种类型的属有蒲公英属（蒲公英）、山柳菊属（山柳菊）和悬钩子属（黑莓等）。

聚合的 加在一个物种名称前面表示在一个密切相关的群体中包含了其他类群。

聚合果 一类由花上数枚离生心皮聚合而成的果实，如悬钩子属（蔷薇科）的果实。[257]

聚生种，复合种 一组两个或多个密切相关的物种，为方便而被给予一个共享名称，如树莓（黑莓）。

翼，翅，翼瓣 豆科蝶形花亚科植物花的两个侧生花瓣之一。[230]

腋下的，腋的（见"腋生的"）。

腋花 生于歧伞花序两分枝之间的花，如石竹科某些属的植物。[204]

具翅的 具一个或多个翅。[159, 173, 177, 258]

胚乳 种子内储存的营养物质。

具胚乳的 拥有胚乳的。

alien Not native to the region concerned.

alkaline soils Soils with a pH value above 7.5.

allele, allelomorph Any one of the alternative forms of a particular gene.

allogamy (cross-fertilisation) Fertilisation of the ovules of a flower by pollen from a different flower. (see **geitonogamy** and **xenogamy**)

allopatric Of plant species or populations, not growing in the same geographical area.

allopolyploid A polyploid of hybrid origin, containing sets of chromosomes from two or more different species.

alternate Placed singly along the stem or axis, not opposite or whorled. [186]

alternation of generations In the life cycle of ferns and fern allies, the alternation of a haploid gametophyte generation, reproducing sexually, with a diploid sporophyte generation, reproducing asexually. [275]

ament (amentum, catkin) A spicate, often pendulous inflorescence of unisexual, apetalous flowers. [204]

amphicarpic, amphicarpous Producing two kinds of fruit, differing in one or more characters.

amphidiploid An allopolyploid containing a diploid set of chromosomes from each of two different species.

amphiphloic siphonostele (solenostele) A type of stele in which a central core of pith is surrounded first by a ring of phloem, then by a ring of xylem, followed by a second ring of phloem. [280]

amphistomatal, amphistomatic With stomata on both upper and lower surfaces of the leaf.

amphitropous (hemitropous) Curved, so that both ends of the ovule are brought near to each other. [223]

外来的 非本地产的，非关注地区产的。

碱性土 pH值在7.5以上的土壤。

等位基因 任何一个特定基因的替代形式。

异花受精 一朵花上的胚珠被来自不同花上的花粉受精（见"同株异花受精"和"异株异花受精"）。

异域种的分布区不重叠的 植物物种或居群不生长在同一地理区域内的。

异源多倍体 杂交起源的多倍体，含来自两个或两个以上不同物种的数套染色体。

互生的 沿茎或轴单生的，非对生的或轮生的。[186]

世代交替 在蕨类和拟蕨类生活史中，单倍体的配子体世代（有性世代）与二倍体的孢子体世代（无性世代）的交替。[275]

柔荑花序 由单性、无瓣花构成的穗状、通常柔软下垂的花序。[204]

具两种果实的 产生两种类型的果实，有一项或多项特征不同。

双二倍体 包含来自两个不同物种之一的二倍体染色体组的异源多倍体。

双韧维管柱 一类在中心髓外首先由一环韧皮部包围，然后由一环木质部包围，再由第二环韧皮部包围的维管柱。[280]

两面气孔的 叶子的上下表面均具有气孔的。

横生的 弯曲，使胚珠的两端彼此靠近。[223]

amplexicaul Clasping the stem, but not completely encircling it. **[187]**

amyloplast A form of leucoplast occurring in storage organs that can convert sugar into starch.

anastomosing Having the veins branched, the vein branches sometimes meeting only at or near the margin of the leaf. **[189]**

anastomosis A cross-connection of veins in a leaf, producing a somewhat denser network of veins towards the margin. **[189]**

anatomy The science or study of the structure of plants, based on dissection.

anatropous With the body of the ovule inverted so that it lies alongside the funicle. **[223]**

ancipitous Having two edges and being flattened, as the pseudobulbs of *Laelia rubescens*.

androdioecious A species in which individual plants bear only male flowers or only bisexual flowers.

androecium The male sex organs (stamens) collectively. **[214—218]**

androgynophore A stalk bearing both androecium and gynoecium, as in the flowers of many members of the Passifloraceae. **[228]**

andromonoecious Having male and bisexual flowers on the same plant.

androphore A stalk bearing the androecium, as in the flowers of some members of the Tiliaceae.

anemochore A plant whose seeds or fruits are dispersed by the wind.

anemochorous Of seeds or fruits, dispersed by the wind.

anemochory The dispersal of seeds or fruits by the wind.

anemophilous Depending on the wind to convey pollen for fertilisation.

抱茎的 抱着茎干，但并不完全环绕它。[187]

造粉体 贮藏器官中白色体的一种形式，可以将糖类转化为淀粉。

网结的 具有分支叶脉的，叶脉分支有时只在叶缘或接近叶缘处相汇。[189]

网结 叶脉交叉连接向叶边缘产生有点密集的网络。[189]

解剖学 基于解剖的关于植物结构的科学或研究。

倒生的 胚珠倒置以致其位于珠柄旁边。[223]

两边的，两棱的 扁平具两棱，像红花蕾丽亚兰的假鳞茎。

雄花两性花异株的，雄全异株的 单一植株上仅有雄花或仅有两性花的物种。

雄蕊群 全体雄性器官（雄蕊）。[214—218]

雌雄蕊柄 着生雄蕊群和雌蕊群的柄，如西番莲科多数植物的花。[228]

雄全同株的，雄花两性花同株的 同一植株上具有雄花和两性花的。

雄蕊柄 着生雄蕊群的柄，诸如椴树科一些植物的花。

风子植物，风播植物 一类靠风传播种子或果实的植物。

风播的 由风传播种子或果实的。

风播 由风传播种子或果实。

风媒的 依靠风传播花粉进行受精的。

anemophily Pollination by means of the wind.

Angiospermae (Anthophyta, Magnoliophyta) The angiosperms, flowering plants whose ovules are enclosed in an ovary. **[213]**

angustiseptate Having the partition (septum) across the narrowest diameter of the fruit, as in *Capsella bursa-pastoris* (Shepherd's Purse). **[264]**

anisophyllous With two leaves of a pair differing in shape or size.

anisophylly The condition of being anisophyllous.

annual A plant that completes its life cycle within a single year.

annual ring A growth ring, formed in the course of a year in the stem or root of a woody plant, that consists of a band of large xylem cells produced in the spring (see **spring wood**), followed by progressively smaller cells produced in the late summer and autumn (see **autumn wood**). **[171]**

annular Ring-like.

annulate Composed of rings or having that appearance. **[147]**

annulus The specialised ring of cells on a sporangium which is involved in the release of the spores. **[275]**

antenna one of the pair of slender structures on the pollinarium of the genus *Catasetum* that, when touched by an insect, cause the pollinia to be forcibly ejected. **[252]**

antepetalous (see **antipetalous**).

antesepalous (see **antisepalous**).

anterior Front, away from the axis. **[226]**

anthela The panicle in some species of *Juncus* (rush), in which the upper branches are overtopped by the lower ones.

anther The part of the stamen that produces pollen. **[212, 216]**

anther cell (see **theca**).

风媒 通过风进行传粉。

被子植物（有花植物、木兰植物门） 被子植物，胚珠被包在子房内的有花植物 。[213]

果实狭隔的 通过果实最窄径的隔膜，如荠菜的果实。[264]

不等叶的，异叶的 一对在形状或大小上具有差异的两片叶子。

不等叶性 处于不等叶的状态。

一年生植物 在一年内完成其生命周期的植物。

年轮 木本植物茎或根在一年过程中形成的生长轮，由春季产生的大量木质部大细胞（见"春材"），和逐步在夏末和秋季产生的小细胞（见"秋材"）组成 。[171]

环状的 环形的。

环状的 由环组成的或外表为环形的。[147]

环带，环形物 这是孢子囊上专门参与释放孢子的一环细胞。[275]

触角 飘唇兰属植物花粉器上的一对纤细结构之一，当其被昆虫碰触时，花粉块被强行弹出。[252]

对瓣的 （见"对瓣的"）。

对萼的 （见"对萼的"）。

前面的，前端的，前部的 前面，远离轴。[226]

长侧枝聚伞花序 某些灯心草属植物灯心草的、下面侧枝超过上面侧枝的圆锥花序。

花药 雄蕊产生花粉的部分。[212, 216]

药室 （见"花粉囊"）。

antheridium The male sex organ in ferns that produces antherozoids. **[275]**

antheriferous Bearing anthers.

antherozoid (spermatozoid) A male sex cell with sets of flagella that enable it to move in water. **[272]**

anthesis (efflorescence) Flowering time. **[244]**

anthocarp A structure comprising a fruit enclosed in a persistent perianth, as in the Nyctaginaceae. **[208]**

anthocyanins The pigments present in solution in the vacuoles of plant cells that are responsible for the red, blue, or purple colouring in flowers, fruits, and other parts of flowering plants. (see also **carotenoids**)

anthophore An extension of the receptacle above the calyx that appears as a short stalk bearing the corolla, stamens, and ovary. **[208]**

Anthophyta (see **Angiospermae**).

antidromous Having the stipules joined by their outer margins, as in *Alchemilla mollis* (Rosaceae). **[178]**

antipetalous Of stamens, situated on the same radii as the corolla segments, as distinct from alternating with them. **[214]**

antipodal cell One of the group of usually three cells, typically haploid, that lie in the embryo sac at the opposite end to the micropyle. **[213]**

antisepalous Of stamens, situated on the same radii as the calyx segments, as distinct from alternating with them. **[214]**

antrorse Pointing forwards or upwards.

aperturate Having pores. **[212]**

aperture An opening, often circular. **[212]**

apetalous Without petals.

apex (plural **apices**) The tip of an organ. **[151, 195]**

apical At the apex of an organ. **[173]**

精子器 蕨类植物产生游动精子的雄性器官。 [275]

具花药的 生有花药。

游动精子 具鞭毛、能在水中游动的雄性生殖细胞。 [272]

花期（开花期） 开花时间。 [244]

掺花果、假果 像紫茉莉科植物一样，果实包在宿存花被内形成的一种结构。 [208]

花色素苷 存在于植物细胞液泡内细胞液中的可使花、果实或植物其他部分染成红色、蓝色或紫色的色素（也见"类胡萝卜素"）。

花冠柄 表现为顶端着生着花冠、雄蕊和子房，由花萼上面的花托延伸来的短柄。 [208]

有花植物 （见"被子植物"）。

异向旋转的 托叶加入其外层边缘，如羽衣草（蔷薇科）。 [178]

对瓣的 雄蕊的特点，不同于其他的相互交替，雄蕊位于与花冠相对的位置上。 [214]

反足细胞 一类通常三个一组的细胞，通常为单倍体，位于胚囊中，在珠孔的另一端。 [213]

对萼的 雄蕊的特点，不同于与其互生，雄蕊位于与花萼裂片相对的位置上。 [214]

顺向的，向上的 向前或向上的。

有孔的 具孔。 [212]

孔 开口，通常圆形。 [212]

无花瓣的 没有花瓣。

顶端 器官的顶端。 [151, 195]

顶端的 在器官的顶端。 [173]

apical placentation (pendulous placentation) The arrangement in which the placenta is situated at the top of the ovary and the ovule or ovules hang down from it. **[223]**

apiculate With an apiculus. **[195]**

apiculus A short sharp point. **[195]**

apocarpous (dialycarpic) Having free carpels. **[219]**

apocarpy The condition of being apocarpous. **[219]**

apogamy Asexual reproduction in ferns, in which a sporophyte is produced directly from a prothallus without the union of gametes.

apomictic Reproducing either by seeds produced asexually, as many species of *Taraxacum* (dandelion) and *Hieracium* (hawkweed), or vegetatively, as when a part of a plant may become detached and develop into a separate plant without any sexual reproduction having taken place.

apomixis Reproduction without fertilisation, either vegetatively (see **vegetative apomixis**) or by seed (see **agamospermy**).

apopetalous (choripetalous, dialypetalous, polypetalous) With a corolla of separate petals, as *Geranium*. **[232]**

apophysis The part of a cone-scale that remains exposed when the cone is closed. **[270]**

aposepalous (chorisepalous, dialysepalous, polysepalous) With a calyx of separate sepals, as *Geranium*. **[232]**

apospory The development of a gametophyte from a sporophyte without the production of spores. **[277]**

appendage An attached subsidiary part.

appressed (adpressed) Lying flat against. **[170]**

aquatic Living in water or a waterlogged environment. **[175]**

arachnoid hairs Fine, interlaced hairs

顶生胎座（悬垂胎座） 胎座类型之一，胎座位于子房的顶端，胚珠从胎座悬垂于子房内。[223]

具细尖的 具有细尖。[195]

细尖 小而细长的尖。[195]

离生心皮的 具离生心皮的。[219]

离生心皮 处于离生心皮的状态。[219]

无配子生殖 蕨类植物的无性生殖方式，孢子体不经过配子结合，直接从原叶体中产生。

无融合生殖的 既可以通过无性繁殖产生的种子来进行，如蒲公英属（蒲公英）和山柳菊属山柳葡的多数种类，也可以是营养繁殖，在没有任何有性生殖发生的情况下，植物体的一部分脱离，发育成单独植物。

无融合生殖、无配生殖 不经过融合，通过营养体（见"营养无融合生殖"）或通过种子（见"无配子种子生殖"）繁殖。

离瓣的（离瓣的、多瓣的） 具有离生花瓣的花冠，如老颧草属植物。[232]

鳞盾 果鳞在球果未张开时露在外面的部分。[270]

离萼的〔离萼（片）的、多萼（片）的〕 具有离生萼片的花萼，如老颧草属植物。[232]

无孢子生殖 配子体直接由孢子体发育而成，不产生孢子。[277]

附属物 一个附带的附属部分。

紧贴的（紧靠的） 平贴在。[170]

水生的 生活在水中或水淹的环境中。[175]

蛛丝状毛 类似于蜘蛛网的精细、交织的

resembling a spider's web, like those on the leaves of *Sempervivum arachnoideum* (Houseleek). **[202]**

毛，像蛛网长生草叶子上的毛。[202]

arboreous, arborescent Tree-like in growth or general appearance.

乔木状的 生长或外观上乔木状的。

archegonium The female sex organ in ferns that produces the egg. **[275]**

颈卵器 蕨类植物产生卵细胞的雌性器官。[275]

arching, arcuate Bending over, curved.

弧曲的 弯腰弯曲的。

areole One of the small areas surrounded by veins in a leaf with reticulate venation **[278]**; one of the small, spine-bearing areas on the stem of a cactus. **[240]**

网眼，刺座 具网状脉序的叶子上被叶脉围成的小区域之一 [278]；仙人掌植物茎上的具刺小区域之一。[240]

aril An outgrowth of the funicle, forming an appendage or outer covering of a seed, e.g. the fleshy, scarlet outer covering of the seeds of *Taxus* (yew). **[159]**

假种皮 珠柄的生长物，能够形成种子的附属物或外壳，如红豆杉属植物（学杉）种子外面的肉质、鲜红色外壳。[159]

arillate With an aril. **[159, 256]**

具假种皮的 有假种皮。[159, 256]

arilloid Resembling an aril.

拟假种皮的 类似假种皮的。

arista An awn or stiff bristle. **[195]**

芒 芒或刚毛。[195]

aristate With an arista. **[195]**

具芒的 有芒的。[195]

aroid A member of the Araceae, as *Arum maculatum* (Lords-and-Ladies). **[207]**

天南星科的 天南星科的成员，如斑叶疆南星。[207]

articulate, articulated Jointed. **[278]**

有节的，分节的 具关节的。[278]

arundinaceous Reed-like.

芦苇状的 芦苇状的。

ascending Sloping or curving upwards. **[162, 164]**

上升的 向上倾斜或弯曲。[162, 164]

ascidiate Bearing pitcher-shaped structures, as the leaves of *Nepenthes*. **[181]**

具瓶状体的 具有瓶状结构，如猪笼草属植物的叶子。[181]

ascidium A little pitcher.

瓶状体 小瓶子。

asepalous Without sepals.

无萼片的 没有萼片。

aseptate Without partitions.

无隔膜的 没有分隔。

asexual Not sexual, i.e. not involving the fusion of male and female cells.

无性的 无性，即不涉及雄性及雌性细胞的融合。

assimilatory Capable of converting inorganic substances into the constituents of the plant system.

同化的 能够将无机物质转化成植物系统成分的。

assurgent Rising upwards.

上升的 上升的。

asymmetric With one side of the leaf larger than the other **[194]**; having flowers not divisible into equal halves, as the flowers of the genus *Canna*. **[236]**

不对称的 叶子的一边大于另一边 [194]；花不能分成相等的两部分，如美人蕉属植物的花。[236]

atactostele A type of stele characteristic of monocotyledons, in which the individual vascular bundles are distributed throughout the ground tissue. **[168]**

atropous (see **orthotropous**)

attenuate Drawn out and gradually narrowing. **[194]**

atypical Not conforming to type.

auct. (auctorum) Of authors, used after a botanical plant name to indicate that that particular name has been accepted by various authors, but not by the original one.

auct.non A phrase placed after a botanical plant name to signify that the name has been misapplied. It is followed by the name of the original author.

auricle A small lobe or ear-like appendage. **[241]**

auriculate With one or more auricles. **[194]**

authority The author(s) of a plant name, i.e. the person(s) responsible for giving a name to a particular taxon. These personal names are normally used in scientific descriptions of plants, on plant labels, etc. for added precision. They are often abbreviated, as in Papaveraceae Juss. (=Jussieu), *Papaver* L. (=Linnaeus), and *Papaver glaucum* Boiss. & Hausskn. (=Boissier and Haussknecht). Certain changes in the classification of a plant may result in a name with a 'double authority' as in *Lobularia maritima* (L.) Desv. In this case, the plant (Sweet Alison or Sweet Alyssum) was originally named *Alyssum maritimum* by Linnaeus, but was subsequently transferred to the genus *Lobularia* by Desvaux.

autochory Dispersal of seeds by the plant itself, e.g. by means of an explosive or ejective mechanism. **[232]**

autogamy (self-fertilisation) Fertilisation of the ovules of a flower by pollen from the

散生中柱 单子叶植物特有的中柱类型，该类型的维管束分布于整个基本组织中 。[168]

（**胚珠**）**直生的** （见"直生的"）。

渐狭 延长并逐渐变窄。[194]

非模式的 不符合模式的。

作者的 用于植物名称后面，表示该特定名称曾被不同作者接受，但不是最初者。

错误鉴定 放在植物名称后面的一个短语、表示该名称曾被错误应用。它后面跟原始作者名称。

耳状物 一个小裂片状或耳朵状附属物。[241]

耳状的 具有一个或多个耳状物。[194]

作者 植物名称的作者，即负责给这个特定类群命名的人。为增加准确度，这些人名通常被用在植物科学描述、植物标签上等。它们经常为缩写形式，如罂粟科Papaveraceae Juss.(=Jussieu)，罂粟属*Papaver* L.(=Linnaeus)，和郁金香罂粟*Papaver glaucum* Boiss. & Hausskn.(=Boissier和Haussknecht)。植物分类上的一些改变可以导致一个名称，如*Lobularia maritima* (L.) Desv.，具有"双重作者"。在这种情况下，植物（香雪球）最初被林奈（Linnaeus）命名为*Alyssum maritimum*，但接下来又被德沃（Desvaux）转移到*Lobularia*（属）。

自体散播 依靠植物体自身，如通过弹裂或喷出方式传播种子。[232]

自花受精（自体受精） 一朵花的胚珠与同一朵花的花粉进行受精作用。

same flower.

autopolyploid A polyploid containing three or more sets of chromosomes, all from the same species.

autumnal Occurring in autumn.

autumn wood (late wood, summer wood) Wood, darker in colour and with smaller xylem cells than spring wood, that is produced in late summer and autumn. **[171]**

auxin One of a group of hormone-like substances, formed in the actively growing parts of plants, that control the growth and development of the plant.

awl-shaped (see **subulate**)

awn A bristle-like appendage, often occurring on the glumes or lemmas of grasses **[243]**; one of the linear structures projecting from the base of the anther in some species of the genus *Erica* **[216]**. the strip of tissue attached to a mericarp in the Geraniaceae. **[232]**

axil The angle formed by the upper side of a leaf and the stem. **[188]**

axile placentation The arrangement in which the placentas are situated on the central axis of the ovary in the angles formed by the septa. **[222]**

axillary (alar) In the axil. **[173, 178]**

axis (plural **axes**) A central line of symmetry in a plant or part of a plant, e.g.stem, root, or rachis. **[267, 270]**

baccate Berry-like.

bacciferous Bearing berries.

back-cross A cross between a hybrid plant and one of its parents.

baculate Of pollen grains, covered with rods that are higher than wide and not constricted at their bases.

baculiform Rod-like.

balausta A many-celled, many-seeded,

同源多倍体 多倍体含三套或三套以上的染色体，都来自同一物种。

秋天的，秋生的 发生于秋季。

秋材（晚材、夏材） 木质颜色更黑，具有比春材更小的木质部细胞，产生于夏末和秋季。[171]

生长素，一组激素状物质的之一，产生于植物生长活跃的部分，控制植物的生长和发育。

锥形的 （见"锥形的"）。

芒 刚毛状的附属物，通常生于禾本科植物的颖片或稃片上 [243]；某些石南属植物花药基部突起的线性结构之一 [216]；附着于牻牛儿苗科植物中果皮上的带状组织。[232]

腋 由叶的近轴面和茎构成的角。[188]

中轴胎座 胎座类型之一。胎座沿隔膜形成的角排列，并位于子房的中心轴线上。[222]

腋生的 在腋内。[173, 178]

轴 植物体或植物体各部分的中央对称线，如茎、根或中轴。[267, 270]

浆果状的 浆果状的。

具浆果的 生有浆果的。

回交 一杂种植物与其亲本之一进行的杂交。

棒的，棒状纹饰的 花粉粒被高长于宽、基部不收缩的棒覆盖。

杆状的，棒状的 棒状的。

石榴果 一类许多室、许多种子、具粗糙果

indehiscent fruit with a tough pericarp, as *Punica granatum* (Pomegranate). **[262]**

band (see **retinaculum**).

banner (see **vexillum**).

barb A hooked hair. **[258]**

barbed With hooked hairs. **[258]**

bark A collective term for all the tissues outside the cambium of a woody stem. **[171]**

basal At the base of an organ. **[168, 194]**

basal placentation The arrangement in which the placenta is situated at the bottom of the ovary. **[223]**

basal plate The 'disc' or reduced stem at the base of a bulb. **[151]**

base The part of attachment of any organ. **[216, 243]**

basifixed With the anther attached by the base to the filament. **[216]**

basipetal Produced, developed, or opening from apex to base.

basiscopic On the side towards the base.

basitonic A type of branching in which the shoots nearest the base of the stem show the greatest development.

bast (see **phloem**).

bathyphyll One of the first or basal fronds of a climbing fern. **[277]**

beak The slender projection from the apex of certain fruits. (see also **rostrum**) **[235, 255, 264]**

beard The line of dense hairs at the base of the outer perianth segments ('falls') in flowers of the genus *Iris*. **[238]**

Beltian body A structure formed at the end of a leaflet in certain plants, e.g. *Acacia cornigera* (Bull's horn Acacia), that is used as food by ants. **[181]**

berry A fleshy, indehiscent fruit with the seed or seeds immersed in pulp. **[207]**

betalain A red or yellow pigment found only in plants of the order Caryophyllales, e.g.

皮的不开裂果实，像石榴的果实。[262]

带 （见"花粉块柄"）。

旗瓣 （见"旗瓣"）。

倒刺 一种具钩的毛。[258]

具倒刺的 具有倒钩的毛。[258]

树皮 一木质茎形成层外所有组织的集合名词。[171]

基部 在器官的基部。[168, 194]

基生胎座，基底胎座 胎座位于子房基部的排列方式。[223]

基盘，鳞茎盘 鳞茎基部的盘状物或退化茎。[151]

基部 任一器官的附着部分。[216, 243]

基着的 花药通过其基部着生于花丝上的。[216]

向基的 从顶端向基部产生、发育或开放。

下侧的 朝向基部的一侧。

单轴分枝的 最接近茎基部的分枝显现最大发育优势的分枝类型。

韧皮部 （见"韧皮部"）。

基生叶 攀缘蕨类的第一枚或基部叶。[277]

喙 某些果实顶端的细长突起。（又见"喙"）[235, 255, 264]

髯毛 鸢尾属植物花的位于外轮花被裂片基部的一行密毛。[238]

贝尔特体 某些植物如金合欢小叶末端形成的、被蚂蚁用作食物的一种结构。[181]

浆果 一类肉质不开裂、具一或数枚埋于果肉中的种子的果实。[207]

甜菜拉因 一类仅在石竹目植物如甜菜中发现的红色或黄色色素。

Beta vulgaris (Beet).

biauriculate Having two auricles.

bicarinate Having two keels.

biciliate With two cilia.

bicollateral bundle A vascular bundle having phloem on two sides of the xylem.

bicoloured Of two colours.

biconvex (see **lenticular**)

bidentate Having two teeth; (leaves) with the margin composed of larger and smaller teeth. **[193]**

biennial A plant that completes its life cycle within two years, producing only vegetative growth in the first year, and flowering in the second.

bifid Divided up to about half-way into two parts. **[201]**

bifurcate Forked, divided into two more or less equal branches. **[220]**

bigeneric Composed of two different genera, as the orchid × *Laeliocattleya*, a hybrid genus produced by crossing a species of *Laelia* with one of *Cattleya*.

bijugate Of a compound leaf, having two pairs of leaflets. **[248]**

bilabiate With two lips, as the corolla in many members of the Labiatae.

bilaterally symmetric (see **zygomorphic**)

bilobed With two lobes. **[208, 271]**

bilocular Having two loculi or compartments.

binomial The botanical name of a plant, comprising the name of a genus followed by the name of the species.

binomial or **binominal nomenclature** The system devised by the Swedish botanist Linnaeus and published in his Species Plantarum (1753), in which plants are distinguished by a two-word name, the first word being the name of the genus (generic

具双耳的 具两个耳状物的。

双龙骨瓣的 具两个龙骨瓣的。

双纤毛的 具两条纤毛的。

双韧维管束 一类韧皮部位于木质部内外两侧的维管束。

二色的 两种颜色的。

双凸的 （见"透镜状的"）。

具双齿的，具重锯齿的 具两齿的；（叶）具由大齿、小齿组成的边缘。[193]

二年生植物 在两年内完成生活史，第一年仅进行营养生长，第二年开花的植物。

二裂的 分裂到近中部成两部分。[201]

二叉的 分叉的，分成两个或多或少等长的分枝。[220]

属间杂交的 由两个不同的属组成的，像兰花蕾丽亚嘉德利亚兰属（×*Laeliocattleya*），是由蕾丽亚属（*Laelia*）植物的种和嘉德利亚属（*cattleya*）植物的种杂交产生的一个杂交属。

具两对小叶的 具两对小叶的复叶的。[248]

二唇形的 具有两枚唇瓣的，像唇形科多数植物的花冠。

两侧对称的 （见"两侧对称的"）。

两裂片的 具两枚裂片的。[208, 271]

具二室的 具有两个室或空间。

双名的 植物的植物学命名，包括属的名称和其后跟的种的名称。

双名法或双名命名法 这个系统是瑞典植物学家林奈创立的，并发表于他的《植物种志》（1753）一书中，植物以两词名称来区分，第一词是属的名称（属名），第二词是种的名称（种加词）。

name) and the second the name of the species (specific epithet).

biochemical Involved with a chemical process in a living organism.

生物化学的 生物有机体中涉及化学过程的。

biovulate Having two ovules.

具双胚珠的 具两枚胚珠的。

bipartite Divided almost to the base into two parts.

二深裂的 几乎裂到基部成两部分的。

bipinnate Pinnate, with the primary leaflets again pinnate. [192]

二回羽状的 羽状复叶，初生小叶再次形成羽状复叶。[192]

bipinnatifid Pinnately lobed, with the lobes themselves similarly divided. [276]

二回羽状分裂的 羽状全裂的，裂片本身再次羽状全裂。[276]

biseriate In two series, rows or whorls. [264]

双列的 两列、两排或两轮的。[264]

biserrate With a saw-toothed margin composed of larger and smaller teeth. [193]

重锯齿的 具由大和小两种锯齿组成的锯齿状边缘的。[193]

bisexual (hermaphrodite, monoclinous) Having both stamens and pistils in the same flower.

两性的（雌雄同体的，雌雄同花的） 一朵花中同时具雄蕊和雌蕊。

bisulcate Of a pollen-grain, having two grooves or furrows.

两沟的 花粉粒具两条槽或沟痕的。

bitegmic Of an ovule, having two integuments.

双珠被的 胚珠具两层珠被的。

biternate Consisting of three parts, each part again divided into three. [192]

二回三出的 包括三部分，每部分又分为三部分的。[192]

bivalved With two valves. [208]

两瓣的 具两枚果瓣的。[208]

blade (see **lamina**).

叶片，瓣片 （见"叶片，瓣片"）。

blind Of a flower bud, failing to develop into a flower.

盲芽的 未能发育成一朵花的花芽的。

bloom The waxy, often bluish green covering on some leaves and fruits; a flower.

蜡被，花 某些叶和果实上的蜡质的、通常蓝绿色的覆盖物；花。

blossom A flower, especially one on a fruit tree; the mass of flowers on a fruit tree.

花 花，特别是果树的花；果树上大量的花。

bole The trunk of a tree. [162]

树干 树的躯干。[162]

boll The spherical or ovoid capsule of *Gossypium* (cotton). [160]

棉桃，棉铃 棉属植物（棉花）的球形或卵形蒴果。[160]

bostryx (helicoid cyme) A spiral inflorescence, with axes on different planes, branching always in the same direction. [206]

螺状聚伞花序 一类轴在不同平面上的螺旋状花序，分枝总在相同方向上。[206]

bough One of the main branches of a tree.

（树的）主枝 树的主要分枝之一。

bowed Curved, arched. [189]

弯曲的 弯曲的，弧形的。[189]

bract A much-reduced leaf, especially the small or scale-1ike leaves associated with a flower or flower cluster. [184, 207]

苞片 极端退化的叶，特别是与花或花簇相关的小的或鳞片状叶。[184, 207]

bracteate Bearing bracts.

bracteolar Relating to bracteoles.

bracteolate Bearing small bracts.

bracteole A small bract, especially when borne on the pedicel of a flower, usually one in monocotyledons, and two (often opposite but sometimes staggered) in dicotyledons. In some species of the genus *Erica* there are three bracteoles, whorled or almost so. **[184]**

bractlet A small bract.

branch A division or subdivision of an axis.

branchlet A small branch or twig. **[273]**

breathing root (see **pneumatophore**)

bristle A stiff hair. **[232]**

bromeliad A member of the Bromeliaceae.

bud A young shoot, protected by scale leaves, from which either leaves or flowers may develop. **[150, 151, 170]**

bud break The stage in the development of a bud when leaves become visible at its apex. **[170]**

bud scale One of the scales that enclose a bud. **[170]**

bud sport A branch, inflorescence, or flower that differs genetically from the remainder of the plant, the differences persisting when the plant is vegetatively propagated from the part concerned.

bulb A usually underground organ, consisting of a short disc-1ike stem bearing fleshy scale leaves and one or more buds, often enclosed in protective scales. **[151, 168]**

bulbous Having or resembling a bulb.

bulbil A small bulb, often one that arises from the axil of a leaf or the inflorescence. **[155]**

bulblet A small bulb.

bullate With blister-like swellings on the surface. **[199]**

bulliform cell (hinge cell) One of the cells that lie in the grooves on the upper surface of the leaves of grasses, and participate in

具苞片的 生有苞片的。

小苞片的 与小苞片相关的。

具小苞片的 生有小苞片的。

小苞片 小苞片，尤其生在花梗上的，单子叶植物通常一枚，双子叶植物两枚（通常对生，但有时不稳定）。某些石南属植物具三枚轮生或近轮生的小苞片。[184]

小苞片 小苞片。

枝 轴上的分枝或再次分枝。

小枝 小枝或嫩枝。[273]

呼吸根 （见"呼吸根"）。

刚毛 坚硬的毛。[232]

凤梨科植物 凤梨科的成员。

芽 被鳞叶保护的、可发育为叶或花的嫩枝条。[150, 151, 170]

芽裂 叶在芽顶端逐渐显现的芽发育阶段。[170]

芽鳞 包住芽的鳞片之一。[170]

芽突变 遗传上与植物其余部分不同的枝、花序或花，当植物用该部分进行营养生殖时这种差异持续存在。

鳞茎 一类通常被包在保护性鳞片中、包含着生着肉质鳞叶和一或多枚芽的短盘状茎的地下器官。[151, 168]

鳞茎的 具鳞茎的或类似于鳞茎的。

珠芽 小鳞茎，通常从叶腋或花序腋内生出。[155]

小鳞茎 小的鳞茎。

具泡状隆起的 表面上具水泡状膨胀物的。[199]

泡状细胞（铰合细胞） 一种位于禾本科植物叶片上表面沟内的细胞之一，参与叶的折叠和展开。[241]

the folding and unfolding of the leaf. [241]

bur, burr A rough or prickly fruit of a plant, aiding dispersal of its seeds by animals; the excrescence on the trunk of a tree formed by the bases of epicormic shoots, as in *Tilia* (lime). [165]

bursicle The flap-like or pouch-like base of the rostellum in some members of the Orchidaceae. [251]

buttress root An adventitious root that grows out from the lower part of the trunk of a tree and remains connected to it down to ground-level. [147]

C (in a floral formula) Corolla, e.g. C5 indicates a corolla composed of 5 petals.

c.(circa) About, approximately.

CaCO₃ (see **calcium carbonate**).

caducous Falling off early.

caespitose (cespitose) Tufted. [165]

calathidium The head of flowers in members of the Compositae, or, in a narrower sense, the involucre only.

calcarate Spurred.

calcicole Growing on soils containing lime.

calcifuge Growing on lime-free soils.

calciphile A plant that prefers soils containing lime.

calcium carbonate (CaCO₃) Lime, chalk.

caliciform Resembling a calyx.

calloused Having a callus. [173]

callus A hard or tough tissue that develops over a wound after an injury [173], or occurs naturally on the labellum of some orchids. [253]

calycanthemous Having sepals wholly or partially converted into petals.

calycle, calyculus A small calyx, or a calyx-like structure often composed of bracts.

calyptra A cap-1ike structure covering

刺果，树瘤 植物的有助于动物传播种子的粗糙或带刺果实；树干上由嫩条枝芽基部形成的树瘤，如椴树属植物（椴树）。[165]

粘囊 兰科某些植物蕊喙的扁平状或袋状基部。[251]

板状根 树干下部生出的、并与其保持连贯一直到达地面的不定根。[147]

花冠（在花程式中） 花冠，如C5表示花冠由5枚花瓣组成。

大约 大约，近似。

碳酸钙 （见"碳酸钙"）。

早落的 过早脱落的。

丛生的 生长密集的。[165]

头状花序，总苞 菊科植物的头状花序，或狭义上仅指总苞。

具距的 有距的。

钙生植物的 在石灰质土壤上生长的。

嫌钙植物的，避钙植物的 在不含石灰质土壤中生长的。

适钙植物，喜钙植物 适合生长在石灰质土壤上生长的植物。

碳酸钙 石灰石，白垩。

花萼状的 类似于花萼的。

胼胝体的 具胼胝体的。[173]

胼胝体 由受伤的木质部发育而来的[173]，或自然发生于某些兰花唇瓣上的坚硬粗糙的组织。[253]

花萼瓣化的 具全部或部分转化成花瓣状的萼片。

副萼 一类通常由苞片组成的小花萼或花萼状结构。

帽状体 覆盖在一些花或果实上的帽状结构，

some flowers and fruits, as the calyx of *Eschscholzia californica* (Californian Poppy). **[231]**

像花菱草的花萼。[231]

calyptrate Bearing or resembling a calyptra. **[231]**

具帽状体的 生有或类似于帽状体的。[231]

calyx (plural **calyces**) The outer perianth, composed of free or united sepals. **[159, 210]**

花萼 由离生或合生萼片组成的外轮花被。[159, 210]

calyx lobe One of the free parts that is joined to the tube of a gamosepalous calyx. **[229]**

花萼裂片 结合成合瓣花花萼萼筒的离生部分之一。[229]

calyx segment (see **sepal**)

花萼裂片 （见"萼片"）。

calyx tube The tube of a gamosepalous calyx. **[160]**

萼筒 合瓣花花萼的筒状部位。[160]

cambial Relating to cambium.

形成层的 与形成层相关的。

cambium (see **vascular cambium** and **phellogen**)

形成层 （见"维管形成层"和"木栓形成层"）。

campanulate Bell-shaped. **[211]**

钟状的 钟形的。[211]

campylotropous With the body of the ovule curved so that it appears to be attached by its side to the funicle. **[223]**

（胚珠）弯生的 胚珠体弯曲以致它显示出其侧面被附着到珠柄上。[223]

canaliculate Channelled.

具沟的 具有沟的。

candelabriform Candelabra-like, with tiered Whorls of radiating branches, as the inflorescence in some species of *Primula*. **[209]**

烛台状的，叠生星状 烛台状的，具有分层轮生的辐射分枝，如某些报春花属植物的花序。[209]

cane The hollow, jointed, woody stem of certain grasses, e.g. *Saccharum officinarum* (Sugar Cane) and species of bamboo; the solid stem of certain species of palm, e.g. *Calamus*, used for making rattan and Malacca canes; the stem of species of *Rubus* (Raspberry, Blackberry, etc.).

茎秆 某些禾本科植物如甘蔗和竹类植物的空心、具节的木质茎；某些棕榈科植物如省藤属植物的实心茎，用于制作藤杖和马六甲手杖；悬钩子属植物（树莓、黑莓等）的茎。

canescent Densely covered with short, greyish white hairs. **[201]**

被灰白色毛的 密被短、灰白色毛的。[201]

capillary Hair-like.

毛状的 毛发状的。

capitate Pin-headed, as the stigma in the genus Primula **[221]**; growing in heads, as the flowers of the Compositae.

头状的，头状花序的 钉头状的，像报春花属植物的柱头 [221]；生长成头状花序的，像菊科植物的花。

capitellate Diminutive of capitate.

小头状的， 小头状花序的。

capitular Relating to a capitulum.

头状花序的 头状花序相关的。

capitulum A head of sessile or almost sessile flowers surrounded by an involucre, the

头状花序 由被总苞围绕的无柄或几乎无柄的花组成的头状花序，尤其是菊科和川

inflorescence especially characteristic of the Compositae and Dipsacaceae. **[205]**

capsular Relating to, or in the form of a capsule.

capsule A dry, dehiscent fruit formed from a syncarpous ovary. **[207, 208, 219, 245, 256]**

capsuliferous Bearing capsules.

carbohydrate A compound based on carbon, hydrogen and oxygen, e.g. sugar, starch, and cellulose.

carbon dioxide (CO₂) A colourless, odourless gas that is necessary for photosynthesis to take place.

carina (keel) The structure formed by the two more or less united lower petals in the flowers of plants of subfamily Papilionoideae in the Leguminosae. **[230]**

carinate Keel-shaped. **[188]**

carnivorous plant (insectivorous plant) One of some 400 species in several different families that live in habitats poor in nutrients, and have developed insect traps of various kinds in order to obtain the nourishment required for their continued existence. **[181]**

carnose, carnous Fleshy, pulpy.

carotenes A group of red or orange pigments, belonging to the carotenoids, that occur in the chromoplasts of plant cells. They are found in the roots of carrots, and in some flowers and fruits.

carotenoids Red, orang, or yellow pigments, including carotenes and xanthophylls, that occur in the chromoplasts of plant cells and often act as accessory photosynthetic pigments. (see also **anthocyanins**)

carpel One of the units forming the gynoecium, usually consisting of ovary, style, and stigma. **[207, 219]**

carpellate (see **pistillate**).

carpophore The stalk which bears the

续断科植物花序的特征。[205]

蒴果的 关于蒴果的，或以蒴果形式的。

蒴果 由合生心皮的子房形成的干燥、开裂的果实。[207, 208, 219, 245, 256]

具蒴果的 生有蒴果的。

碳水化合物 一类基于碳、氢和氧的化合物，如糖类、淀粉和纤维素。

二氧化碳 一种光合作用进行时必需的无色、无味气体。

龙骨瓣（龙骨状突起，龙骨瓣） 豆科蝶形花亚科植物花中由两个多少联合的下花瓣构成的结构。[230]

龙骨瓣状的 龙骨状隆起的。[188]

肉食植物（食虫植物） 数个不同科的约400种植物之一，它生活在营养匮乏的生境，为获取继续生存所需的营养，发育成了各种捕虫器。[181]

肉质的 肉质的、多汁的。

胡萝卜素 一类发生于植物细胞有色体中、属于类胡萝卜素的红色或橘黄色的色素。它们见于胡萝卜的根和一些植物花、果实中。

类胡萝卜素 一类发生于植物细胞有色体中、通常作为辅助性光和色素、包括胡萝卜素和叶黄素的红色、橘黄色或黄色的色素（也见"花色素苷"）。

心皮 形成雌蕊的单位之一，雌蕊通常包括子房、花柱和柱头。[207, 219]

具心皮的 （见"具雌蕊的"）。

心皮柄，果瓣柄 石竹科蝇子草属和蝇子草

fruit in the genus *Silene* and some other members of subfamily Silenoideae in the Caryophyllaceae **[208]**, also in the family Umbelliferae. **[257]**

cartilaginous Gristly.

caruncle A protuberance near the hilum of a seed. **[157]**

carunculate With a caruncle. **[157]**

carunculoid Resembling a caruncle.

caryopsis (plural **caryopses**) A dry, one-seeded, indehiscent fruit, characteristic of grasses, having the pericarp united to the seed; the grain of a cereal grass. **[158, 244]**

casual An alien plant that has not become naturalised.

catalyst A substance that increases the rate of a reaction without itself being changed, e.g. an enzyme.

cataphyll A reduced leaf, e.g. a bract, bracteole, bud scale, or one of the papery, sheathing leaves which enclose the whole of the newly developing aerial shoot in the genus *Crocus*. **[150]**

catkin (see **ament**)

caudate With a tail-like appendage. **[195]**

caudex The stem of a plant, especially a fern or a woody monocotyledon, e.g. a palm. **[247, 280]**

caudicle The flexible, stalk-like group of threads connecting a pollinium with the viscidium in the Orchidaceae. **[152]**

caulescent Having an obvious stem. **[165]**

cauliflorous Exhibiting cauliflory. **[208]**

cauliflory The production of flowers on the trunk and branches of trees rather than at the ends of twigs, as in *Cercis siliquastrum* (Judas Tree) and *Theobroma cacao* (Cocoa Tree). **[208]** (see also **ramiflory**)

cauline Borne on the stem. **[168, 169]**

cell The smallest unit of plant tissue that can function independently, containing a nucleus

亚科一些植物着生果实的柄 [208]，也见于伞形科植物。[257]

软骨质的 软骨的，似软骨的。

种阜 种子种脐附近的突起。[157]

具种阜的 具有种阜。[157]

种阜状的 类似于种阜的。

颖果 一种干燥、含一粒种子、果皮不裂的果实，禾本科植物特有的果实类型，具与种子合生的果皮；禾谷植物的谷粒。[158, 244]

非归化外来植物 没有被归化的外来植物。

催化剂 一种自身不被改变、只加快反应速率的物质，如酶。

低出叶 一类退化叶，如苞片、小苞片、芽鳞、或番红花属植物包住整个新生地上枝的纸质、鞘状叶之一。[150]

柔荑花序（见"柔荑花序"）。

尾状的 具一尾状附属物。[195]

茎 植物的茎，特别是蕨类或木本单子叶植物，如棕榈。[247, 280]

花粉块柄 兰科植物连接花粉块和黏盘的一束柔韧、柄状的丝状物。[152]

具茎的 具明显茎的。[165]

茎生花的 展示茎生花现象的。[208]

茎生花现象 花生长在树干、树枝上，而不是生长在嫩枝顶端，像在紫荆和可可树中。[208]（又见"枝生花现象"）

茎生的 生于茎上的。[168, 169]

细胞、室 植物组织的最小独立的功能单位，含有细胞核和其周围的细胞质

surrounded by cytoplasm **[173, 196]**; a compartment of an anther or an ovary. (see **loculus** and **theca**)

cellular Relating to a cell.

censer mechanism A form of seed dispersal occurring in genera such as *Papaver* (poppy) and *Antirrhinum* in which the capsules open in such a way that seeds are scattered only when the capsule is swung from side to side, as in a strong wind.

centrifugal Developing from the centre towards the margin.

centripetal developing from the margin towards the centre.

cephalium A woody enlargement, bearing a dense mass of hairs, at or near the top of the stem in certain cacti. **[240]**

cereal Any grass that produces an edible grain (caryopsis).

cespitose (see **caespitose**).

cf., cfr.(confer) Compare, used in a similar way to the abbreviation 'aff.'.

chaff Thin, dry, membranous bracts or scales, especially the bracts at the base of the florets in members of the Compositae. **[234]**

chaffy Chaff-like.

chalaza The basal portion of the nucellus of an ovule. **[223]**

chalazal Relating to a chalaza.

chalazogamous With the pollen tube entering the ovule through the chalaza.

chalazogamy The entry of the pollen tube into the ovule through the chalaza.

chamaephyte A small, woody or herbaceous perennial, having resting buds not more than 25 cm above soil level. **[166]**

chartaceous Paper-like.

chasmogamous With flowers opening normally for reproduction to take place.

chasmogamy The production of flowers which open in the normal way to expose the

[173, 196]；花药或子房的内部空间（见"室"和"药室"）。

细胞的、室的 与细胞或室有关的。

摇动机制 发生在如罂粟属（罂粟）和金鱼草属植物的种子传播的一种方式，蒴果开裂，只有随风左右摇动时种子才被散出。

离心式发育的 从中心向边缘逐渐发育的。

向心式发育的 从边缘向中心逐渐发育的。

花座 某些仙人掌植物茎顶端、生有密集毛团的木质膨大物。 [240]

谷类植物 任一能产生食用谷粒（颖果）的禾草类植物。

丛生的 （见"丛生的"）。

比较 比较、用法类似于缩写"aff."。

膜片 薄、干燥、膜质的苞片或鳞片，尤其是菊科植物小花基部的苞片。 [234]

膜片状的 形似膜片。

合点 胚珠的珠心基部。 [223]

合点的 与合点有关的。

合点受精的 花粉管经过合点进入胚珠的。

合点受精 花粉管经过合点进入胚珠。

地上芽植物 一类小的、具土壤表面以上不超过25厘米高的休眠芽的木本或多年生草本植物。 [166]

纸质的 像纸的，纸状的。

开花受精的 以正常开放使生殖得以进行的花的。

开花受精 以普通方式开放来暴露生殖器官的花的结果。

reproductive organs.

chasmophyte A plant that grows in rock crevices.

chemonasty A nastic movement in response to a chemical stimulus.

chimaera, chimera A plant or part of a plant composed of cells of two genetically different types, either by a mutation or by the grafting together of parts from two different individuals. (see **graft chimaera**)

chiropterophilous Pollinated by bats.

chiropterophily Pollination by bats.

chlorenchyma Parenchymatous tissue containing chloroplasts, e.g.the mesophyll in a leaf. [196, 268]

chlorophyll The green pigment in plants which allows photosynthesis to take place.

chlorophyllous Containing chlorophyll.

chloroplast A plastid containing the green pigment chlorophyll necessary for photosynthesis. **[196]**

chlorosis An unhealthy condition due to a deficiency of chlorophyll that causes the green parts of the plant to become yellowish.

choripetalous (see **apopetalous**)

chorisepalous (see **aposepalous**)

chromoplast A plastid containing a pigment, especially the red, orange or yellow pigments known as carotenoids.

chromosome One of the pairs of strands in the nucleus of a cell that bears genes in a linear order. The number of chromosomes in a cell will vary according to the species, cultivar, hybrid etc. concerned.

ciliate Fringed with long hairs. [193]

ciliolate Fringed with short hairs.

cilium One of the fine hairs, resembling eyelashes, that arise from the margin of an organ. [193]

cincinnus (scorpioid cyme, spiralled cyme) A cylindrical inflorescence, with axes on

石隙植物 生长于岩石裂缝中的植物。

感药性，趋化性 响应化学刺激物的感性运动。

嵌合体 由两种遗传上不同类型的细胞经突变或两不同个体的部分嫁接在一起组成的植物或植物的一部分（见"嫁接嵌合体"）。

蝙蝠媒的 蝙蝠传粉的。

蝙蝠媒 通过蝙蝠传粉。

绿色组织 含有叶绿体的薄壁组织，例如叶内的叶肉。 [196, 268]

叶绿素 植物体内允许光合作用进行的绿色色素。

叶绿素的 含叶绿素的。

叶绿体 一种含有光合作用所必需的绿色色素叶绿素的质体。 [196]

黄化 一种因叶绿素缺乏导致植物绿色部分变黄的不健康状况。

离瓣的 （见"离瓣的"）。

离萼的 （见"离萼的"）。

有色体，色质体 一类含色素，尤其被称为类胡萝卜素的红色、橘黄色或黄色色素的质体。

染色体 细胞核中携带线状排列的基因的成对链之一，细胞中染色体数目将根据相关种、品种、杂种等发生变异。

具长缘毛的 边缘具长毛的。 [193]

具短缘毛的 边缘具短毛的。

缘毛，纤毛 从器官边缘生出的、类似于睫毛的细毛之一。 [193]

蝎尾状聚伞花序（蝎尾状聚伞花序，螺旋状聚伞花序） 圆柱形的花序，花轴在

different planes, branching alternately to one side and the other. **[206]**

circinate Coiled in a flat spiral, like a young fern frond. **[188]**

circumfloral Around the flower.

circumscissile Dehiscing by a line round the fruit or anther, the top coming off as a lid. **[256]**

circumscissile capsule (see **pyxis**).

cirrhiferous Bearing tendrils. **[179]**

cirrhose, cirrhous, cirrose Ending in a long, coiled tip, tendril-like. **[195]**

cl. Clone.

clade A group of plants believed to have evolved from a common ancestor.

cladistic Relating to a clade.

cladistics A method of classification based on the assumed divergence of groups of plants from a common ancestor.

cladode (cladophyll, phylloclade) A branch taking on the form and functions of a leaf, as in some members of the Euphorbiaceae and the genus *Ruscus* in Liliaceae. **[181, 240, 268]**

cladogenesis The formation of a species by evolutionary divergence from an ancestral species.

cladogram A tree diagram representing the relationships between species or groups of species based on the cladistic method of classification.

cladophyll (see **cladode**).

class A taxonomic rank standing between division and order. Names of classes end in '-opsida'.

classification The arrangement of plants in increasingly specialised categories because of similarities in their structure.

clathrate Resembling lattice-work.

clavate Club-shaped, thickened towards the apex. **[268]**

不同的平面上，分枝成左右间隔生出。[206]（编者按，解释与图不符）

拳卷的 扁平螺旋状卷曲的，像幼蕨的叶。[188]

环绕花的 花周围的。

周裂的 沿环绕果实和花药周围的一条线开裂的，顶部像盖子一样脱落。[256]

周裂蒴果 （见"盖果"）。

具卷须的 生有卷须的。[179]

卷须状的 （枝、叶等）止于一长而螺旋状顶端的，卷须状的。[195]

无性系 无性系。

进化支 被认为是从一共同祖先进化而来的一类植物。

进化支的 与进化支相关的。

分支系统学 基于一假定共同祖先的植物类群分歧的分类方法。

叶状枝 承担叶形状和功能的分枝，如大戟科和百合科假叶树属某些植物。[181, 240, 268]

系枝发生，分枝进化 从一个古老种经过进化分歧而形成种的过程。

分枝图，进化树 一种表示基于分枝分类方法的种或种组间关系的树状图。

叶状枝 （见"叶状枝"）。

纲 位于门和目之间的分类等级。纲的名称以"-opsida"结尾。

分类 植物在因其结构相似性而逐渐增多的专门等级中的排列方式。

粗筛孔状 类似于网格的。

棍棒状的 棍棒状的，向顶部变更粗。[268]

clavellate Diminutive of clavate.

claw The narrowed base of some petals. **[227]**

cleft Deeply divided into two lobes at the apex. **[195]**

cleistogamous With self-pollination occurring within the unopened flower. **[208]**

cleistogamy The production of often inconspicuous flowers that do not open, allowing only self-pollination to take place, as in some members of the Violaceae. **[208]**

climber A plant that grows upwards by twining round nearby plants and other supports, or by clinging to them with tendrils. **[162, 177]**

clinandrium Of orchids, the tissue at the apex of the column lying beneath the anther.

cline A gradual variation in one or more characters within a species or population.

clip (see **corpusculum**).

clockwise twining Looked at from above, the tip of the climbing plant grows in the direction of the hands of a clock, forming an apparently left-hand spiral, as in *Lonicera* and a few other plants. **[162]**

clone (cl.) A group of plants that have arisen by vegetative reproduction from a single parent and which are therefore genetically identical.

cluster An indeterminate inflorescence containing several flowers. **[208]**

clustered Closely arranged in a group. **[170, 186]**

cm Centimetre(s).

CO₂ (see **carbon dioxide**).

coalescence Union.

coalescent Having grown together.

cob The infructescence of *Zea mays* (Maize, Sweet Corn) consisting of a woody stem to which the grains are attached. **[158, 244]**

cochleate Spirally twisted, like a snail shell.

coconut fibre (see **coir**).

小棍棒状的 小棍棒状的。

爪 某些花瓣的狭窄基部。[227]

半裂的 在顶部深裂为两个裂片。[195]

闭花受精的 自花传粉发生在不开放花内的。[208]

闭花受精 不开放、仅允许自花传粉发生的不显著花的结果。如堇菜科的某些种类。[208]

攀缘植物 以缠绕在附近植物或其他支撑物上，或以卷须附着于附近植物或其他支撑物上向上生长的植物。[162, 177]

药窝 兰花合蕊柱顶端的、位于花药下面的组织。

梯度变异 一个种或居群内一个或更多特征的逐渐变异。

着粉腺 （见"着粉腺"）。

顺时针方向缠绕 从上面看，缠绕植物的顶端以钟表指针的旋转方向生长，形成明显的左螺旋，如忍冬属和少数其他植物。[162]

无性系 从单一亲本经营养繁殖产生的、遗传上相同的一群植物。

花簇 一类含有数朵花的无限花序。[208]

丛生的 紧密排列成一组的。[170, 186]

厘米 厘米。

二氧化碳 （见"二氧化碳"）。

愈合 结合，联合。

愈合的 已经生长在一起的。

玉米穗 玉米（玉米、甜玉米）的果序，包括一个木质茎及附着其上的谷粒。[158, 244]

螺旋状的 螺旋状卷曲的，像蜗牛壳。

椰壳纤维 （见"椰壳纤维"）。

coenocarpium A fruit formed from an entire inflorescence, such as *Ficus* (fig) or *Ananas* (pineapple). **[262]**

coenocarpous Bearing a coenocarpium.

coensorus An extended sorus or a group of sori that have combined so as to appear as one. **[279]**

co-florescence A lateral branch of a synflorescence.

coherent In close contact with a similar part, but not fused with it. **[217]**

coir A fibrous material, processed from the husks of *Cocos nucifera* (Coconut), that is used in the manufacture of mats, ropes and brushes, and also horticulturally as a substitute for peat.

coleoptile The sheath that surrounds the plumule in grasses, and protects it as it grows to the surface of the soil. **[158]**

coleorhiza The sheath that surrounds the radicle in grasses. **[158]**

collateral Lying or standing side by side.

collective fruit (see **multiple fruit**).

collective species (see **aggregate species**)

collenchyma A specialised form of parenchyma, with thickening in the cell walls, that acts as supporting tissue in leaf veins, petioles, and the cortex of stems.

colleter One of the multicellular, glandular hairs that are found on the adaxial surface of sepals, stipules, the base of petioles, and the adjacent surface of stems in e.g. Loganiaceae and Rubiaceae. **[202]**

colpate Furrowed. **[212]**

colporate Having a composite aperture, consisting of a furrow and a pore. **[212]**

colpus An elongated aperture that appears as a furrow on the surface of a pollen grain. **[212]**

columella The persistent central axis round which the carpels of some fruits are arranged.

column The combination of stamens

聚花果 由整个花序发育形成的果实，如无花果属植物（无花果）和凤梨属凤梨植物。[262]

聚花果的 生有聚花果的。

聚合孢子囊群 一个伸展的孢子囊群或结合成一体的一群孢子囊群。[279]

侧枝花序 复合花序的一枚侧枝。

粘贴的 与相似的部分紧密相接，但不与之融合。[217]

椰壳纤维 一类从椰子外壳加工成的纤维材料，用于制造衬垫、绳索和毛刷，园艺上也被用作泥炭的替代品。

胚芽鞘 禾草类植物围绕胚芽、并保护胚芽生长至土壤表面的鞘。[158]

胚根鞘 禾草类植物围绕胚根的鞘。[158]

并生的 肩并肩卧着和立着的。

聚花果 （见"聚花果"）。

复合种 （见"复合种"）。

厚角组织 一种具细胞壁加厚、作为叶脉、叶柄和茎皮层支持组织的薄壁组织的特化形式。

黏液毛 在如马钱科和茜草科植物萼片、托叶近轴面、叶柄基部和茎的邻近表面发现的多细胞腺毛之一。[202]

（萌发）沟的 沟状的。[212]

（萌发）孔沟 具由沟和孔构成的综合孔的。[212]

（萌发）沟 花粉粒表面一种以沟的形式出现的长孔。[212]

中轴，果轴 某些果实的周围有心皮排列的中心轴。

合蕊柱 雄蕊和花柱结合成的单一结构，

and styles into a single structure, as in Asclepiadaceae and Orchidaceae. **[233, 251]** (see also **rod**)

columnar Growing in the shape of a vertical cylinder, as some conifers. **[164]**

coma The tuft of hairs at the end of some seeds, as those of *Asclepias* (milkweed) and *Gossypium* (cotton) **[159, 160]**; the apical crown of leaves on the fruit of *Ananas comosus* (Pineapple). **[262]**

commensalism A form of symbiosis in which two different organisms co-exist, only one benefiting, but the other not being harmed.

commissure The surface along which adjoining carpels are appressed. **[257]**

comose Having a tuft of hairs.

compatible Of any two plants, able to fertilise each other.

composite A member of the Compositae.

compound Composed of two or more similar parts. **[177, 192, 204, 205]**

compressed Flattened.

compression wood A kind of reaction wood found on the lower sides of branches and inclined trunks of softwood trees and characterised by rounded tracheids with intervening intercellular spaces.

concave Appearing as if hollowed out. **[269]**

concolorous Having the same colour throughout.

conduplicate Of leaves, folded once lengthwise **[188]**; of a style, grooved down one side, giving the appearance of being folded. **[220]**

cone The conical expansion at the base of the beak of the fruit of *Taraxacum* (dandelion). **[235]** (see also **strobile**)

confluent Merging together.

congeneric Belonging to the same genus.

conical Cone-shaped. **[164]**

conifer A plant bearing cones, as *Pinus sylvestris* (Scots Pine). **[267—270]**

Coniferae (see **Pinopsida**).

像萝藦科和兰科。 [233, 251]（也见"柱"）

柱状的 直立柱状生长的，如某些球果植物。[164]

种缨，叶冠 某些种子末端的一簇毛，像马利筋属植物（马利筋）和棉属植物（棉花）的种子 [159, 160]；凤梨果实顶端的叶冠。[262]

偏利共生 两种不同生物共存、仅一方收益但另一方也不受伤害的一种共生现象。

接合面 邻近两个心皮相接的面。[257]

丛毛的 具簇毛的。

亲和的 能彼此受精的任何两种植物的。

菊科植物 菊科的植物。

复合的 由两个或多个相似部分组成的。[177, 192, 204, 205]

平扁的 压扁的。

应压木 一类在树枝和倾斜软木树树干下侧发现的畸形木材，以圆形管胞具明显的细胞间隙为特征。

凹入的 好像被掏空的。[269]

同色的 全部具同一颜色的。

折合状 叶子纵向折叠一次的 [188]；花柱一侧具向下的槽，仿佛被折叠一样的 [220]

喙基 蒲公英属植物（蒲公英）果实喙基部的圆锥状膨大物。[235]（也见"球果"）

汇合的 汇合在一起的。

同属的 属于同一属的。

球果状的 球果状的。[164]

球果类 一类生有球果的植物，如长白松。[267—270]

松柏纲 （见"松柏纲"）。

conifer allies The Cycadopsida, Ginkgoopsida, and Gnetopsida. **[271—273]**

coniferous Bearing cones.

conker The seed of *Aesculus hippocastanum* (Horse-chestnut). **[256]**

connate United with a similar part as stipules **[178]**, bracts **[207]**, stamens **[217]**, or styles. **[220]**

connate-perfoliate With the bases of two opposite leaves joined together so that the stem appears to pass through them. **[187]**

connective The part of the stamen, a continuation of the filament, that joins together the two pairs of anther cells. **[216, 226]**

connivent (convergent) Gradually approaching each other and meeting at the tips.

conoid, conoidal Cone-shaped.

conspecific Belonging to the same species.

contiguous Touching, but not united with.

contorted Twisted **[163]**; of sepals or petals, overlapping at one margin and overlapped at the other.

contractile root A root that can shorten in order to keep a bulb, corm or rhizome at a particular level. **[150—152]**

convar. Convariety.

convariety (convarietas, plural convarietates) A concept developed by the German botanist F. Alefeld (1820—1872) to group together closely related varieties of cultivated plants, e.g. *Brassica oleracea* ssp. *oleracea convar. fruticosa*, which comprises var. *ramosa* (Thousand-headed Kale) and var. *gemmifera* (Brussel Sprout).

convars. Convarieties.

convergent (see **connivent**).

convex Having a rounded surface.

convolute In vernation, having one leaf rolled inside another. **[189]**

cordate Heart-shaped. **[190, 194]**

拟球果类 苏铁纲、银杏纲和买麻藤纲。 [271—273]

具球果的 生有球果的。

七叶树果实 欧洲七叶树的种子。 [256]

合生的 与相似部分如托叶 [178]、苞片 [207]、雄蕊 [217] 或柱头合生的。 [220]

合生穿叶的，对生连基抱茎的 对生叶基部结合在一起以致茎仿佛穿过它一样。 [187]

药隔 雄蕊的部分，将两对药室连在一起的花丝延伸物。 [216, 226]

靠合的 彼此逐渐靠近，至顶端靠合。

球果形的 球果状的。

同种的 属于同一种的。

临接的 接触的，但不与之合生。

扭曲的，旋转状的 （枝条等）扭曲的 [163]；萼片或花瓣的一侧边缘交叠和另一侧缘被交叠的。

收缩根 一类能够变短以此保持一特定程度的鳞茎、球茎或根状茎的根。 [150—152]

品类 品类。

品类 德国植物学家F. Alefeld（1820—1872）提出的、将一些密切相关的栽培植物变种组合在一起的概念，如甘蓝品类包括千头甘蓝和抱子甘蓝。

品类 品类（复数）。

会聚的 （见"靠合的"）。

凸的 具一圆形表面。

席卷的，旋转的 幼叶卷叠式，一片叶被卷到另一叶内面。 [189]

心形的 心形的。 [190, 194]

core The central part, containing the seeds, of the fruit of *Malus* (apple), *Pyrus* (pear) and similar fruits. [261]

coriaceous Leathery.

cork (see **phellem**)

cork cambium (see **phellogen**)

corky Having the outer surface cork-like. [173]

corm A swollen underground stem, somewhat bulb-like in appearance, but solid and not composed of fleshy scale leaves. [150]

cormel (cormlet) A small corm.

corniculate Bearing a small horn or horn-like outgrowth.

corolla The inner perianth, composed of free or united petals. [208, 226]

corolla lobe One of the free parts that is joined to the tube of a gamopetalous corolla. [229]

corolla segment (see **petal**).

corolla tube The tube of a gamopetalous corolla. [229]

corona A structure occurring between (and sometimes united with) the stamens and the corolla, as the cup-shaped or trumpet-shaped outgrowth in the genus *Narcissus*, the horned cuculli in *Asclepias curassavica*, and the ring of filaments in the genus *Passiflora* (passion flower). [228, 233]

coronal scale One of the ring of scales on the inner surface of the corolla, as at the junction of the limb and the claw in *Silene dioica* (Red Campion). [227]

corolloid Corolla-like.

corpus The inner layers of cells in an apical meristem, which divide to produce the inner tissues of the shoot.

corpusculum The clip connecting the two bands (retinacula) which are attached to the pollinia in the Asclepiadaceae. [233]

cortex (plural **cortices**) The outer part of an organ; the tissue in a stem or root between

果核 苹果属植物（苹果）、梨属植物（梨）及类似果实的含有种子的中心部分。[261]

革质的 皮革的。

木栓 （见"木栓"）。

木栓形成层 （见"木栓形成层"）。

木栓的 具木栓状的外表面。[173]

球茎 膨大的地下茎，外表有点鳞茎状，但实心、不由肉质鳞叶组成。[150]

小球茎 一个小的球茎。

具小角状突起的 生有一个小角和小角状的生长物。

花冠 内轮花被，由离生或合生的花瓣组成。[208, 226]

花冠裂片 基部合生成合瓣花筒部的离生裂片之一。[229]

花冠段 （见"花瓣"）。

花冠筒 合瓣花花冠的筒部。[229]

副花冠 生于雄蕊和花冠之间（有时与之合生）的结构，像水仙属植物花的杯状或喇叭状的生长物，马利筋花的角状盔帽和西番莲属植物（西番莲）花的呈轮丝状物。[228, 233]

副鳞 花冠内面上的一轮鳞片之一，像异株蝇子草花瓣的瓣片和爪结合处的鳞片。[227]

花冠状的 花冠状的。

原体 分裂产生枝内部组织的顶端分生组织的内细胞层。

着粉腺 萝藦科植物花的连接两个携带花粉块的花粉块柄的夹状结构。[233]

皮层 器官的外部；茎和根位于表皮和维管组织之间的组织。[148, 168]

the epidermis and the vascular tissue. **[148, 168]**

cortical Relating to the cortex.

corticate Having bark, or a bark-like covering.

corymb A racemose inflorescence with pedicels of different lengths, causing the flower cluster to be flat-topped. **[205]**

corymbiform In the form of a corymb.

corymbose In the form of a corymb; bearing corymbs.

costa A rib, the mid-vein of a simple leaf or the rachis of a compound leaf.

costapalmate With the petiole extending into the blade of a palmate leaf, as in some palms. **[249]**

costate Ribbed.

costule The midrib of a pinnule or of the lobe of a pinna.

cotyledon (seed leaf) One of the first leaves of the embryo of a seed plant, typically one in monocotyledons, two in dicotyledons, and two or more in gymnosperms (up to eighteen in the Pinaceae). **[157—160, 267]**

coumarin An aromatic substance, the smell of newly mown hay, especially of the grass *Anthoxanthum odoratum*.

counter-clockwise twining Looked at from above, the tip of the climbing plant grows in the opposite direction to the hands of a clock, forming an apparently right-hand spiral. Most climbers are in this category. **[162]**

crassinucellate Having a nucellus of considerable bulk, because of the numerous cells lying between the epidermis and the embryo sac.

creeper A plant that grows along the ground, over fences, or up walls, e.g. *Hedera helix* (Ivy).

cremocarp A dry fruit, composed of two one-seeded carpels, that at maturity separate into mericarps, e.g. *Heracleum* (hogweed) and

皮层的 与皮层有关的。

具树皮的 具有树皮，或树皮状覆盖物的。

伞房花序 一类具不同长度花柄、导致形成平顶的总状花序。[205]

伞房花序状的 伞房花序状的。

伞房花序的 伞房花序状的；生有伞房花序的。

中脉 肋，一个单叶的中脉或复叶的叶轴。

具肋掌状的 具伸入掌状叶叶片的叶柄，像某些棕榈类植物。[249]

具中脉的，具脉的 具肋的，具枝的。

小肋 小羽片或羽片裂片的中肋。

子叶 种子植物胚的第一叶，典型地单子叶植物一片子叶，双子叶植物两片，裸子植物中有两片或更多（松科植物达八片）。[157—160, 267]

香豆素 一类具芳香性气味的物质，具新割制的干草特别是黄花茅的气味。

逆时针方向缠绕 从上面看，缠绕植物顶端以逆时针方向生长，形成明显的右螺旋，多数缠绕植物属于这一类。 [162]

具厚珠心的 具一相当体积的珠心，因为多数细胞位于表皮和胚囊之间。

匍匐植物或攀缘植物 一类沿地面、蔓延于栅栏或附着于墙壁上生长的植物，如洋常春藤。

双悬果 一类由两枚含一粒种子的心皮构成的干果，成熟时裂成分果爿，如独活属和其他伞形科植物。[257]

other members of the Umbelliferae. **[257]**

crenate With rounded teeth. **[193]**

crenulate With small, rounded teeth. **[193]**

crest An irregular or toothed ridge on the upper part of an organ **[221, 233]**; in the genus *Iris*, one of the two lobes at the end of each style branch. **[238]**

crested Bearing a crest.

crispate With the margin curled or crumpled. **[193]**

cristate Crested.

croceate Saffron-coloured.

cross-fertilisation (see **allogamy**)

cross-pollination The transfer of pollen from one plant to another.

crown The upper, branched part of a tree above the bole **[162, 163]**; the rootstock of e.g. *Rheum* x *cultorum* (Rhubarb).

crownshaft The cylinder formed by tubular leaf sheaths at the apex of the stem in some species of palm. **[247]**

crozier (fiddlehead) A coiled young fern frond. **[275]**

cruciate (see **cruciform**).

crucifer A member of the Cruciferae, a family in which the flowers typically have four petals, arranged in the form of a cross, and tetradynamous stamens. **[215]**

cruciferous Resembling a crucifer.

cruciform (cruciate) In the form of cross. **[211]**

cryptobiosis Extended dormancy, as when dehydrated seeds and pollen-grains, also certain flowering plants and ferns in dry habitats remain dormant for long periods but rapidly become active again with the arrival of rain.

cryptobiotic Capable of remaining dormant for a long time in the absence of water.

cryptogam (flowerless plant) A plant belonging to the Cryptogamia, a division in

具圆齿的 具圆形齿。[193]

具细圆齿的 具小圆形齿的。[193]

鸡冠状突起 器官上部的一类不规则或牙齿状的纵脊 [221, 233]；在鸢尾属植物中，每一花柱分枝末端的两个裂片之一。[238]

具鸡冠状突起 生有脊或鸡冠状突起的。

皱波状的 具弯曲或褶皱状边缘的。[193]

鸡冠状的 具鸡冠状突起的。

藏红花色的 具藏红花颜色的。

交叉受精 （见"异花受精"）。

异花授粉 花粉从一植株转移到另一植株。

树冠，根茎 树的主干以上的分枝部分 [162, 163]；大黄的根茎。

冠轴 某些棕榈类植物茎顶端由管状叶鞘形成的圆柱状结构。[247]

拳卷幼芽（蕨菜） 卷曲的蕨类植物幼叶。[275]

十字形的 （见"十字形的"）。

十字花科植物 十字花科的一员，该科植物花一般有十字形排列的四个花瓣和四强雄蕊。[215]

十字花科植物的 类似于十字花科植物的。

十字形的 十字形状的。[211]

隐生现象，潜生现象 延长休眠期，如脱水种子和花粉粒，还有干旱生境的某些有花植物和蕨类保持长期休眠，但遇雨水到来即迅速激活。

隐生的，潜生的 在缺水状态下保持长期休眠的能力。

隐花植物（无花植物） 一类属于隐花植物门的植物，在前期分类系统中，该门包

former classifications that included all the plants that do not produce seeds, as ferns, fern allies, mosses, fungi, and algae.

cryptophyte A plant with resting buds lying either beneath the surface of the ground, as on rhizomes or in bulbs, or submerged in water. **[166]**

cucullate Hood-shaped.

cucullus A corona hood in the Asclepiadaceae. **[233]**

culm The jointed stem, especially the flowering stem of grasses. **[241]**

cultigen One of many plants found only in cultivation, including cultivars, many hybrids, also a number of ornamental and crop plants which have been grown and developed by man for so long that their wild origins are now uncertain or have been lost entirely.

cultivar (cv., cvar.) A cultivated variety, not necessarily attached to a single botanical species, that has been selected for a particular attribute or combination of attributes. Its name is printed in Roman type with a capital initial letter, and enclosed in single quotation marks, e.g. *Prunus avium* 'Plena', the double-flowered form of *Prunus avium* (Wild Cherry).

cuneate Wedge-shaped. **[191, 194]**

cup (see **cupule**).

cupular Relating to or shaped like a cupule. **[184, 248]**

cupulate Bearing a cupule. **[184]**

cupule A cup-shaped structure composed of coalescent bracts, as in the Fagaceae. In e.g. *Quercus* (oak) the cupule (acorn) is indehiscent, but in *Castanea* (chestnut) the cupule splits into several valves. **[184]**

cupuliferous Having a cup-shaped structure surrounding the nut.

cupuliform Shaped like a cupule.

括所有不产生种子的植物，像蕨类、拟蕨类、苔藓、菌类和藻类。

隐芽植物 一类具休眠芽的植物，休眠芽既可在地面以下，如生于根状茎或鳞茎上，也有沉于水下的。[166]

兜状的 兜形的。

兜瓣 萝藦科植物的兜状副花冠。[233]

秆 具节的茎，尤其是禾草类植物开花的茎花葶。[241]

分类学来源不详植物 仅在栽培过程中发现的许多植物之一，包括品种、许多杂种以及大量的观赏植物和农作物，它们经人类长期培育、生长发育以致其野生起源现已不确定或已完全消失。

品种 栽培变种，不必要附着于单一植物学种，是因一特殊属性或数个属性的组合而被选育出来的。它的名字以具大写首字母的罗马体印刷，并括在单引号内。例如重瓣甜樱桃，甜樱桃的重瓣类型。

楔形的 楔形的。[191, 194]

壳斗 （见"壳斗"）。

壳斗的，壳斗状的 与壳斗相关的，或形状像壳斗的。[184, 248]

具壳斗的 生有壳斗的。[184]

壳斗 一类由合生苞片组成的杯状结构，像壳斗科植物。如栎属植物（橡树）的壳斗不开裂，但栗属植物（栗）的壳斗裂成数瓣。[184]

具壳斗的 具围绕坚果的杯状结构的。

壳斗状的 形状像壳斗的。

cushion The part of the prothallus of a fern that bears the archegonia, often thicker than the surrounding area. **[275]**

cuspidate Ending rather abruptly sharp point. **[195]**

cuticle The waxy layer of cutin that covers the outer surface of the epidermis and restricts the passage of water and gases into and out of the plant. **[196, 268]**

cutin A mixture of fatty substances that comprise the cuticle.

cv., cvar. Cultivar.

cvs., cvars. Cultivars.

cyathiform Cup or beaker-shaped.

cyathium The inflorescence of the genus *Euphorbia*. **[207]**

Cycadopsida The class that comprises the cycads. **[272]** (see **Gymnospermae**)

cyclic Arranged in whorls.

cylindric, cylindrical Shaped like a straight tube, but completely solid.

cymbiform (see **navicular**)

cyme A branching, determinate inflorescence, with a flower at the end of each branch. **[204]**

cymose In the form of a cyme; bearing cymes.

cymule A small cyme or portion of one.

cypsela A small, dry, one-seeded, indehiscent fruit formed from two united carpels, as in the Compositae. Sometimes treated as a form of achene. **[235]**

cyst A sac or bladder-like structure, containing a liquid secretion.

cystolith A stone-like mass, usually of calcium carbonate, that forms within the epidermal cells of certain plants. **[197]**

cytoplasm The protoplasm of a plant cell excluding the nucleus. **[212]**

daughter bulb A small bulb that is produced by a mature (mother) bulb. **[168]**

垫状物 蕨类原叶体上生有颈卵器的部分，通常厚于周围区域。 [275]

骤尖的 末端突然成一锐尖。 [195]

角质层 覆盖在表皮外面的苍白色的角质层，限制水分和气体进出植物体。 [196, 268]

角质 一类组成角质层的脂肪类物质的混合物。

品种 品种。

品种 品种（复数）。

杯状的 杯状的或高脚杯状的。

杯状聚伞花序或杯状花序 大戟属植物的花序。 [207]

苏铁纲 含有苏铁植物的一个纲 [272]（见"裸子植物"）。

轮列的 轮状排列的。

圆柱状的 形状像一根直的管子，但完全实心的。

舟形的 （见"舟状的"）。

聚伞花序 一类分枝的有限花序，每一分枝顶端生一花。 [204]

聚伞花序状的 聚伞花序状的；生有聚伞花序的。

小聚伞花序 小的聚伞花序或聚伞花序的一部分。

连萼瘦果 一类由两合生心皮的子房发育成的、小、干燥、含一枚种子、不开裂的果实，如在菊科植物中。有时被处理为瘦果的一种形式。 [235]

分泌腔 内含液体分泌物的囊状或膀胱状结构。

钟乳体 某些植物表皮细胞内形成的、成分常为碳酸钙的石头状团块。 [197]

细胞质 植物细胞内排除细胞核的原生质体 。 [212]

子鳞茎 成熟（母）鳞茎上产生的小鳞茎 。 [168]

daughter cell One of the two cells produced when a cell divides.

子细胞 一个细胞分裂时产生的两个细胞之一。

decandrous Having ten stamens.

十枚雄蕊的 具有十枚雄蕊。

decaploid (10n) Having ten sets of chromosomes in each cell.

十倍体（10n） 每个细胞含有十套染色体。

deciduous Falling off, as leaves in autumn or petals after flowering time. [173]

脱落的 像秋季的叶子或花期后的花瓣一样落下。[173]

declinate Arching downwards, then turning up towards the apex.

下倾的 弓状下行，再弯向顶端。

decumbent Lying along the ground, but with the tip ascending. [162]

外倾的，匍生的 匍匐于地面，但端部上升。[162]

decurrent Running down, as when the base of a leaf is prolonged down the stem as a wing. [187, 220]

下延的 下延的，如叶基部像翅一样向下延伸到茎。[187, 220]

decussate In opposite pairs, each pair at right angles to the next. [186]

交互对生的 对生成对的，每对与下一对成直角。[186]

deflexed Bent abruptly downwards.

外折的 突然向下弯曲的。

degenerate To deteriorate, to lose normal qualities. [173]

退化的 失去正常品质的。[173]

dehisce Open spontaneously when ripe.

开裂 成熟时自发开裂。

dehiscence The process of splitting open at maturity, usually applied to an anther shedding pollen or a fruit releasing seeds. [218, 219]

开裂 成熟时裂开的过程，通常指花药散出花粉或果实释放种子。[218, 219]

dehiscent Opening naturally.

开裂的 自然开裂的。

deliquescent Rapidly becoming semi-liquid.

溶解的 迅速成为半液态的。

deltoid Triangular. [190]

三角形的 三角形的。[190]

deme A group of plants in a particular locality forming a subpopulation of a species, capable of interbreeding freely amongst themselves but genetically distinct from neighbouring plants of the same species.

同类群 在特定地区的一群植物，能够形成一种植物亚居群，它们之间能够自由杂交，但遗传上不同于同一物种的邻近植物。

dendriform Tree-like.

树状的 树状的。

dendritic Of hairs, branched like a tree. [202]

树枝状的 分枝像树状的毛的。[202]

dendrochronology The study of the annual rings present in trees to determine their age and also to ascertain variations in the climatic conditions during their period of growth; often used in the case of ancient timber to date past events.

树木年代学 一种确定树木年龄、弄清生长期内气候变化的树木年轮的研究；通常用于研究古木来确定古代事件的情形。

dendrogram A tree diagram reflecting the

系统树，树状图 一种用来表达植物物种或

relationships of species or groups of plants, e.g. cladogram.

dendrology The study of trees.

dentate Toothed. [193]

denticulate Minutely toothed. [193]

depauperate In an impoverished condition through lack of the essential elements for growth.

depressed-globose (see **subglobose**).

dermal Relating to the outer covering of an organ.

dermatogen The outermost layer of the apical meristem which develops into the epidermis.

determinate A form of inflorescence in which the terminal flower opens first and prevents further growth of the stem or branch, e.g. a cyme. [204]

diadelphous With two groups of stamens, as in some of the Leguminosae. [215]

dialycarpic (see **apocarpous**).

dialypetalous (see **apopetalous**).

dialysepalous (see **aposepalous**).

diandrous Having two stamens.

diarch root A root with two protoxylem strands in the stele. [149]

diatropic Growing at right angles to the source of the stimulus.

diatropism The growth movement of part of a plant at right angles to the source of the stimulus, as horizontal rhizomes to gravity.

dichasial cyme, dichasium A cyme with lateral branches on both sides of the main axis. [204]

dichlamydeous Having a perianth of two whorls, i.e. both calyx and corolla are present.

dichogamous With stamens and pistils of a flower not ripening at the same time, i.e. either protogynous or protandrous.

dichogamy The state of being dichogamous.

类群之间关系的树状图，如分支图。

树木学 树木的研究。

具牙齿的 具牙齿的。[193]

具细牙齿的 具小牙齿的。[193]

发育不良的 缺乏生长必需元素的贫瘠条件。

扁球形 （见"近球形的"）。

表皮的 与器官最外层有关的。

表皮原 将来发育成表皮的顶端分生组织最外层。

有限的 一种顶生花首先开放、阻止茎或分枝进一步生长的花序类型，如聚伞花序。[204]

二体雄蕊的 具两组雄蕊的，像豆科一些植物。[215]

离生果瓣的 （见"心皮离生的"）。

离瓣的 （见"离瓣的"）。

离萼的 （见"离萼的"）。

两雄蕊的 具两枚雄蕊。

二原型根 一类中柱中具两个原生木质部脊的根。[149]

横向性的 与刺激物源成直角生长的。

横向性 （和外来刺激方向成直角的倾向）植物体部分与刺激物源成直角的生长运动，像重力作用下的水平方向上的根状茎。

二歧聚伞花序 一类主轴两侧具侧枝的聚伞花序。[204]

两被的 花被具有两轮，即有花萼和花冠同时存在。

雌雄异熟的 一朵花的雄蕊和雌蕊（原著为柱头）不在同一时间内成熟，即雌蕊先熟的或雄蕊先熟的。

雌雄异熟 雌雄异熟的状态。

dichotomous Forked, having the axis divided into two equal branches. **[247]**

dichotomous key A device for the identification of plants, composed of couplets of opposing statements (leads), whereby the user successively rejects one of the two statements until the final lead is reached, resulting in the identification of the plant concerned.

dichotomy Division into two Parts.

diclinous (see **unisexual**).

dicliny The state of being diclinous.

dicotyledons (Dicotyledonae, Magnoliopsida) Flowering plants having two cotyledons. **[169]**

dicotyledonous Having two cotyledons.

dictyostele A type of stele, derived from a siphonostele, which is broken up into a network of meristeles. **[280]**

didymous In pairs, as the mericarps forming the fruit of plants in the Umbelliferae. **[257]**

didynamous Having one pair of stamens longer than the other pair, as in many of the Labiatae. **[215]**

diffuse Widely or loosely spreading.

diffuse-parietal With the ovules scattered over the inner surface of a carpel.

digestive zone The area within the pitcher-like leaf of a carnivorous plant where the trapped insects decompose and become the source of nutrients. **[181]**

digitate Palmate with narrow leaflets. **[192]**

dilated Broadened, enlarged.

dimerous Having the parts of the flower in twos.

dimorphic Occurring in two forms.

dioecious Having male and female flowers on different plants of the same species. **[227]**

dioecism, dioecy The state of being dioecious. **[227]**

diploid (2n) Having two sets of chromosomes

二歧的 叉状的，具分成两个相等分枝的轴的。[247]

二歧式检索表 一种植物鉴定的工具，由一对意义相反的陈述句（线索）组成，以此使用者依次拒绝两个陈述句中的一个，直到到达最后线索，最终鉴定出相关植物。

二歧 分成两部分。

雌雄异花的 （见"单性的"）。

雌雄异花 雌雄异花的状态。

双子叶植物 （双子叶植物亚纲，木兰纲）具两枚子叶的有花植物。[169]

双子叶的 具两枚子叶的。

网状中柱 一种从管状中柱演化而来、破裂形成网状分体中柱的中柱类型。[280]

成对的 成对的，像形成伞形科植物果实的分果爿的。[257]

二强雄蕊的 具一对长于另一对雄蕊的，如许多唇形科植物。[215]

散开的 广泛或松散地展开的。

散生侧膜胎座的 胚珠散生于心皮内表面上的。

消化区 肉食植物囊状叶内部、将被困昆虫分解为营养来源的区域。[181]

指状的 掌状具狭窄小叶（似手指）的。[192]

膨大的 扩展的，扩大的。

二基数的 花的各部分以二为基数的。

二型的 以两种形式出现。

雌雄异株的 雄花和雌花生于同种植物的不同植株上。[227]

雌雄异株 雌雄异株的状态。[227]

二倍体的 每一细胞中具两套染色体的。

in each cell.

diplostemonous Havin g the stamens in two whorls, the inner whorl opposite the petals and the outer whorl opposite the sepals. **[214]**

具外轮对萼雄蕊的 具有两轮雄蕊，内轮与花瓣相对，外轮与萼片相对。[214]

diporate Of a pollen grain, having two pores.

双孔的 花粉粒具两个（萌发）孔的。

dipterous Two-winged.

双翅的 具两翅的。

disc A development of the receptacle within the perianth, often bearing nectar glands; the central part of the capitulum in members of the Compositae **[234]**; the basal plate or reduced stem at the base of a bulb. **[151]**

花盘，基盘 花被内面、通常生有蜜腺的花托衍生物；菊科植物头状花序的中心部分 [234]；鳞茎基部的盘状物或退化茎 。[151]

disc floret One of the central, regular flowers in the flower heads of some members of the Compositae. **[175, 234]**

盘状小花 菊科某些植物头状花序中部的整齐花。[175, 234]

disciferous Bearing a disc, often with nectar glands.

具花盘的 生有花盘的，通常具蜜腺。

discoid, discoidal Resembling a disc, i.e. thin, flat, and circular. **[221]**

盘状的 类似于盘子的，即薄、扁平、圆形的。[221]

discrete Separate.

分离的 分开的。

disjunct Discontinuous, with the distribution split into two or more areas.

间断的 不连续的，分布区被分成两个或更多地区的。

disk (see **disc**)

花盘（见"花盘"）。

dispersal The process of scattering seeds away from the parent plant, by means of the wind, birds, animals, etc.

散布；分散 通过风、鸟类、动物等将种子从母株上散布出去的过程。

disperse To scatter, especially in regard to seeds.

散布，传播 散布，特别是关于种子。

dissected Deeply divided into segments.

（叶）多裂的 深裂成裂片的。

distal Situated away from the point of attachment. **[186]**

远基的，远轴的 位于离开附着点的。[186]

distichous Arranged in two vertical ranks. **[186]**

二列的 排成两垂直的列。[186]

distinct Having similar parts completely separate from each other, e.g. as one petal from another.

离生的 具彼此完全分离的相似部分的，如一枚花瓣离生于另一枚花瓣。

distribution The occurrence of a plant species from a geographical point of view.

分布 从地理学观点研究一个植物物种的发生。

distyly Dimorphic heterostyly, the occurrence of two different lengths of style in flowers of the same plant species, as *Primula veris* (Cowslip). **[229]**

花柱两型性，花柱异长现象 同一植物如黄花九轮草的花，具两种不同长度花柱的现象。[229]

disymmetric Divisible through the centre of the flower in only two longitudinal planes for the halves of the flower to be mirror images, e.g. *Dicentra* (Fumariaceae).

dithecous With two anther cells. **[216]**

diurnal Occurring in the daytime.

divaricate Widely spreading or greatly divergent.

divarication Wide-angled branching and subsequent interlacing of the branches of shrubs growing in dry, wind-swept habitats.

divergent Spreading away from each other. **[163]**

divided With lobes extending almost or entirely to the midrib or the base of the leaf.

division (phylum) The highest principal taxonomic rank into which the plant kingdom is divided. Names of divisions end in '-phyta'.

dolabriform Shaped like the head of an axe or a hatchet.

domatium A cavity or tuft of hairs on a plant that acts as a shelter for insects, mites, or other similar organisms. **[182]**

dormancy The state of being inactive.

dormant Resting, inactive. **[150, 170]**

dorsal The side of an organ facing away from the axis, abaxial. **[188]**

dorsifixed With the anther attached by the back to the filament. **[216]**

dorsiventral Having the dorsal and ventral surfaces of the leaf different from each other.

double flower A flower having petals or petal-like structures (petaloids) additional to those in the typical form. Where there is a large number of petals in the double form, the intermediate state may be described as semi-double. **[210]**

downy (see **pubescent**).

drepanium A sickle-shaped cyme, with axes

双对称的 从花中心仅沿两个纵向面将花分成互为镜像的两半，如荷色牡丹属植物（紫堇科）的花。

二室的 （花药）具两药室的。 [216]

昼开性的 在白天发生的。

极叉开的 宽角度开展的或极度叉开的。

极叉开 生于干燥、多风生境的灌木枝条宽角度开展，并彼此交错。

开展的 彼此反向开展的。 [163]

全裂的 裂片几乎或完全延伸至中肋或叶基部的。

（生物分类学上的）门 植物界分类的最高基本分类等级。门的名称以"-phyta"结尾。

斧形的 形似斧头或小斧子。

（叶及植物体上）虫菌穴 植物体上用作昆虫、螨虫和其他类似生物居所的空腔或毛簇。 [182]

休眠 一种不活动的状态。

休眠的 休眠的，不活动的。 [150, 170]

远轴面的，背面的 器官朝向离开轴方向的一面。 [188]

背着的 花药以背面附着于花丝上。 [216]

背腹面的 叶具彼此不同的背、腹面的。

重瓣花 一类在典型类型花基础上，具额外花瓣或花瓣状结构（花瓣状物）的花。重瓣类型具大量花瓣，而中间状态可以描述为半重瓣的。 [210]

具短绒毛的 （见"被短柔毛的"）。

镰状聚伞花序 镰刀状的聚伞花序，轴仅在

only on one plane, branching always to the same side. **[206]**

drip tip The long, narrow apex of the leaves of many plants living in wet, tropical regions that enables water to be rapidly shed from the upper surface. **[181]**

drooping (see **nutant**).

dropper A shoot growing downwards from a bulb or corm and producing a bulb or corm at its apex. **[154]**

drupaceous Resembling a drupe.

drupe A fleshy fruit containing one or more seeds, each enclosed within a stony endocarp, as in species of the genus *Prunus* (Rosaceae). **[259]**

drupelet One of the small drupes which together form an aggregate fruit, as in the genus *Rubus* (Rosaceae). **[257]**

duct Any of the tubular structures forming the vascular tissue which allow the passage of air or water throughout the plant.

duramen (see **heartwood**).

dwarf A plant of much smaller size than the average of its kind; a part of a plant that is undersized in comparison with the average for that kind of part e.g. dwarf shoots. **[267]**

ear A spike or head of a cereal grass, e.g. *Triticum* (wheat) or *Hordeum* (barley), that contains the grains. **[241]**

early wood (see **spring wood**).

ebracteate Without bracts.

ebracteolate Without bracteoles.

ecalcarate Without spurs.

eccentric Off-centre, having the axis not centrally placed. **[220]**

ecdemic Not native.

echinate Bearing spines. **[256]**

echinulate Bearing small spines or bristles.

ecoclimate The climate of a particular habitat.

ecodeme A deme occupying a particular

一个平面上，分枝总朝向同一侧。[206]

滴水叶尖 许多生长在潮湿热带地区的植物的、使水迅速从上面流下来的狭长的叶先端。[181]

下垂的 （见"下垂的"）。

茎出条 从鳞茎或球茎上向下生长的、在其先端再产生鳞茎或球茎的嫩条。[154]

核果状的 类似于核果的。

核果 一类具一枚或多枚种子、每枚种子都包在坚硬内果皮内的肉质果实，如李属（蔷薇科）植物的果实。[259]

小核果 聚合在一起形成聚合果的小核果之一，如悬钩子属（蔷薇科）植物的果实。[257]

输导组织 一类形成维管组织的、允许空气或水分被输导到植物全身各处的任何管状结构。

心材 （见"心材"）。

侏儒植物 一类比其同类植物平均大小小得多的植物；与同类部分平均大小相比小得多的植物部分，如短枝。[267]

谷穗 禾谷植物含有谷粒的花穗，如小麦属植物（小麦）或大麦属植物（大麦）。[241]

早材 （见"春材"）。

无苞片的 没有苞片的。

无小苞片的 没有小苞片的。

无距的 没有距的。

离心的 具离开中心位置的轴的。[220]

外来的 非原生的。

具刺的 生有刺的。[256]

具小刺的 生有小刺或刚毛的。

生态气候 特定生境的气候。

生态同类群 占据特定生态生境的同类群。

ecological habitat.

ecological Relating to ecology.

ecology The study of the plants and animals that occupy a particular habitat, and the interaction between them and their environment.

ecosystem The concept of a community of plants and animals that occupy a particular habitat, with special emphasis on the interaction between them and their environment. The living part of the system is usually divided into 'producers' (green plants). 'consumers' (animals), and 'decomposers' (bacteria and fungi), and the whole process can be represented as a flow diagram.

ecotype The variation within a species, often treated as a subspecies, that occurs in a particular habitat.

ectophloic siphonostele A type of stele in which a central core of pith is surrounded first by a ring of xylem and then by a ring of phloem. **[280]**

edaphic Relating to the soil.

efflorescence (see **anthesis**).

e.g. (exempli gratia) For example.

egg (see **ovum**).

eglandular Without glands.

elaiosome An appendage on the seeds of some plants (e.g. *Helleborus, Viola*) that contains oily substances attractive to ants, which assist in the disperal of the seeds by carrying them away from the parent plant. **[160]**

elater In species of *Equisetum* (horsetail), an elongated appendage attached to a spore which uncurls as it dries and assists in its dispersal. **[283]**

ellipse A figure shaped like a flattened circle. [190]

ellipsoid Elliptical in outline with a three-dimensional body.

生态学的 与生态学有关的。

生态学 研究占据特定生境的植物、动物之间，以及与其环境之间相互作用的科学。

生态系统 占据特定生境的植物、动物的群落，特别强调它们及其与环境之间的相互作用的概念。系统的有生命部分通常分为生产者（绿色植物）、消费者（动物）和分解者（细菌和真菌），整个过程可以用流程图表示。

生态型 占据特定生境的、通常被处理为亚种的种间变异。

外韧管状中柱 一类中心髓的周围首先环绕一轮木质部，然后再环绕一轮韧皮部的中柱。[280]

土壤的 与土壤有关的。

花期 （见"花期"）。

例如 例如。

卵 （见"卵细胞"）。

无腺体的 没有腺体的。

油质体 某些植物（如铁筷子属、堇菜属）的种子附属物，含吸引蚂蚁的脂类物质，从母株上携走种子，有助于其传播。[160]

弹丝 木贼属植物（木贼）附着于孢子上的能够加长的附属物，干燥时能够快速伸直，有助于散布孢子。[283]

椭圆 一种形状像扁圆的图形。[190]

椭球 三维空间体轮廓椭圆形。

elliptic, elliptical In the form of an ellipse. [190]

椭圆形的　椭圆形的。[190]

emarginate Distinctly notched at the apex. [195]

顶端微凹的　顶端具明显凹口的。[195]

embryo A young plant developed sexually or asexually from the ovum. In spermatophytes it is contained within the seed. [157—159, 223]

胚　从卵有性或无性发育成的幼嫩植物体，在种子植物中胚包在种子内。[157—159, 223]

embryonic Relating to an embryo or any other part in a rudimentary state of development.

胚的　与胚或发育起始阶段其他部分相关的。

embryo sac The cell inside the ovule of an angiosperm in which fertilisation occurs, and which develops into the female gametophyte. [213]

胚囊　被子植物胚珠内，能够进行受精，发育成雌配子体的细胞。[213]

em., emend. (emendavit) An abbreviation that indicates an emendation or correction. It is followed by the name of the author responsible for the alteration.

校正，修改　表示校正或修改的缩写。其后面跟负责校改的作者名称。

endemic Restricted to a particular country or region.

特有的　局限于一特定国家或地区的。

endocarp The innermost layer of the pericarp. [260]

内果皮　果皮的最内层。[260]

endodermal Relating to the endodermis.

内皮层的　与内皮层有关的。

endodermis The innermost layer of cells of the cortex in stems and roots. [148, 149, 268]

内皮层　茎和根皮层的最内细胞层。[145, 149, 268]

endogenous Originating or developing from inside the plant.

内起源的　从植物体内部起源或发育的。

endophyllous Formed from within a sheathing leaf.

内生的　从鞘状叶内形成的。

endophyte A plant that lives inside another plant.

内生植物　生于另一植物内的植物。

endorhizal With the radicle, instead of lengthening, giving rise to secondary rootlets, as in monocotyledons. [158]

有鞘幼根的　具代替伸长并生出次生小根的胚根，像单子叶植物。[158]

endosperm The albumen of a seed, particularly that deposited within the embryo sac. [157—159, 260]

胚乳　尤指储存在胚囊中的种子的胚乳。[157—159, 260]

endospermic or **endospermous seed** A seed in which the nutritive tissue is absorbed more slowly into the developing embryo so that part of it remains at least until germination, e.g. *Ricinus communis* (Gastor-oil Plant) and *Zea*

具胚乳种子　一类营养物质更加缓慢地被吸收到正在发育的胚以致其部分至少保留到萌发的种子，如蓖麻和玉米。[157, 158]

mays (Maize, Sweet Corn). **[157, 158]**

endozoic Having seeds or fruits dispersed by animals, e.g. edible fruits, whose seeds are later expelled with the waste material.

endozoochory Dispersal of seeds or fruits by passing unharmed through the digestive system of an animal.

ensiform Sword-shaped. **[190]**

entire With an unbroken margin, i.e. not toothed or lobed. **[193]**

entomophilous Depending on insects to convey pollen for fertilisation.

entomophily Pollination by insects.

enucleate Without a nucleus.

environment The sum total of external influences acting on a plant.

enzyme A protein produced by a living cell which acts as a catalyst for a biochemical reaction.

ephemeral Lasting for a relatively short period.

epiblem, epiblema (see **piliferous layer**).

epicalyx (plural **epicalyces**) A whorl of sepal-like organs Just below the true sepals. **[210]**

epicarp (exocarp) The outermost layer of the pericarp. **[259]**

epicaulic (see **epicormic**)

epichile The apical portion of the labellum in an orchid. **[253]**

epicormic Of shoots, growing out from the trunk of a tree or from a substantial branch. **[165]**

epicotyl The part of a seedling above the cotyledon(s) that gives rise to the stem and leaves. **[157]**

epidermal Relating to the epidermis.

epidermis The outermost layer of cells on plant organs, forming a protective covering to roots, stems, leaves, flowers and fruits. **[148, 168, 169, 171, 196]**

epigeal A type of germination in which the

动物体内传播的 通过动物进行的种子和果实传播。如可食用的果实，其种子后来随废物一起排出。

动物体内传播 不受损害地通过动物的消化系统来进行的种子和果实传播。

剑形的 剑形的。[190]

全缘的 具有不裂的边缘，即不具齿或裂片的。[193]

虫媒传粉的 为受精依靠昆虫传粉的。

虫媒传粉 依靠昆虫进行传粉。

去核的 没有细胞核的。

环境 作用于植物的外界影响的总和。

酶 一类生活细胞产生的、对生物化学反应起催化作用的蛋白质。

短暂的 持续相对短的时期。

根被皮 （见"根毛层"）。

副萼（复数为epicalyces） 真正萼片下面的一轮萼片状器官。[210]

外果皮 果皮的最外层。[259]

嫩条的 （见"嫩条的"）。

上唇 兰花唇瓣的顶端部分。[253]

嫩条的 从树干或实心树枝上生出枝条的。[165]

上胚轴 子叶以上的、生出茎和叶的幼苗部分。[157]

表皮的 与表皮有关的。

表皮 植物器官最外面的形成各器官如根、茎、叶、花和果实等保护层的细胞层。[148, 168, 169, 171, 196]

出土萌发 一类子叶露出地面之上的萌发类

cotyledons emerge above the ground. [157]

epigynous With the sepals, petals and stamens inserted near the top of the ovary. [218]

epinasty The tendency in part of a plant to grow more rapidly on the upper side so that it curves downwards.

epipetalous Borne upon the petals or corolla-segments, as the stamens in many species. [214]

epiphyllous Borne upon the leaves, as the flowers of *Helwingia* (Cornaceae).

epiphylly Growth upon a leaf, as flowers or vegetative buds, e.g. *Kalanchoe daigremontiana* (Crassulaceae). [155]

epiphyte A plant that grows on another plant but does not derive nourishment from it, as many species of tropical orchids, bromeliads and ferns. [146, 182, 277]

epiphytic Growing on another plant, but not deriving nourishment from it. [146, 182, 277]

epistomatal, epistomatic With stomata only on the upper surface of the leaf.

epitepalous Borne upon the tepals.

epithelium A layer of cells, often secretory, that line internal cavities in plants, e.g. resin and gum canals. [158, 268]

epithem A group of water-secreting cells in the mesophyll of the leaves of some plants.

epizoic Having seeds or fruits dispersed by animals, e.g. by being furnished with hooks that become entangled in an animal's fur or hair.

epizoochory Dispersal of seeds or fruits by being carried away on the coats of animals.

equator The surface area of a pollen grain midway between the poles. [212]

equitant Overlapping in two ranks, as the leaves in the genus *Iris*. [186, 189]

erect Upright. [163]

ericaceous Relating to the Ericaceae.

型。[157]

（花被、雄蕊等）上位的 具有着生于子房顶端附近的萼片、花瓣和雄蕊的。[218]

（叶等的）偏上性 植物体某部分上面生长较快致使向下弯曲生长的趋势。

花瓣上着生的 生于花瓣或花冠裂片上的，像许多植物的雄蕊。[214]

叶生的 生于叶上的，像青荚叶属（山茱萸科）植物的花。

叶生现象 生于叶上，如大叶落地生根（景天科）的花和营养芽。[155]

附生植物 生于另一植物上但不从其吸取营养的植物，如许多热带兰科、凤梨科和蕨类植物。[146, 182, 277]

附生的 生于另一植物但不从其吸取营养的。[146, 182, 277]

上表面具气孔的 仅叶上表面具气孔的。

花被着生的 生于花被片上的。

上皮细胞 沿植物内部腔隙排列的、通常为次生性的一层细胞，如树脂或树胶道。[158, 268]

通水组织 某些植物叶肉中一群泌水细胞。

动物体外传播的 依靠动物，如通过种子和果实上的钩状物与动物体表的毛发勾结缠绕而进行的种子和果实传播。

动物体外传播 依靠动物体表带走种子或果实的传播。

赤道面 花粉粒过极轴中部的表面积。[212]

（叶子）基部嵌叠的 叶子呈二纵列交互排列，基部互相包叠，像鸢尾属植物的叶子。[186, 189]

直立的 直立的。[163]

杜鹃花科的 与杜鹃花科植物有关的。

ericetum A heath plant-association dominated by the genus *Erica*.

ericoid Resembling the genus *Erica*.

erose Having an eroded or jagged margin.

escape A plant, previously growing in a garden, that has reproduced and become established outside it.

estipulate Without stipules.

etaerio An aggregate fruit composed of achenes, follicles, berries, or drupes. **[257]**

et al. (et alii) And others, a phrase used after the name of an author to indicate that other authors also were involved in the naming of a particular plant.

etiolated Grown pale and weak through lack of light. **[174]**

etiolation The condition of a green plant growing in insufficient light, resulting in weak stems, long internodes, and small, yellowish or whitish leaves. **[174]**

evanescent Short-lived, not permanent.

even-pinnate (see **paripinnate**).

evergreen Retaining most of its leaves throughout the year.

exalbuminous Lacking endosperm.

exarillate Without an aril.

exceeding Longer than (another organ).

exfoliating Peeling off in layers, flakes or scales, as the bark of *Platanus* (plane).

exfoliation The process of shedding pieces of bark.

exine (extine) The outer part of the wall of a pollen grain, composed of the base, rods or columns, tectum, and ornamentation. **[212]**

exocarp (see **epicarp**).

exodermal Relating to the exodermis.

exodermis The outermost layer of cells in the cortex of a root. **[148]**

exogenous Originating from outside the plant.

exotic Not native; originating in a foreign country, especially one in the tropics.

矮欧石南灌丛 以欧石南属植物占优势的植物灌丛。

似欧石南植物的 似欧石南属植物的。

啮蚀状的 具有侵蚀或使成锯齿状边缘的。

逃逸，逸生 原来生长于公园里的植物已经在园外繁殖并定殖。

无托叶的 没有托叶的。

聚心皮果 由瘦果、蓇葖果、浆果或核果构成的聚合果。 [257]

等，和其他 一个用在一人名后面表示其他作者也参加了这一特定植物命名的短语。

黄化的 因缺少光线而生长变弱、绿色变浅。 [174]

黄化现象 绿色植物因生长于光线不足的环境，而使茎变细弱、节间变长、叶子变小、黄绿色或近白色。 [174]

短暂的 短暂的，不持久的。

偶数羽状的 （见"偶数羽状的"）。

常绿的 在全年内仍然保持绝大多数的叶子。

无胚乳的 缺乏胚乳的。

无假种皮的 无假种皮的。

超过 长于（另一器官）。

片状剥落 像悬铃木属植物（悬铃木）的树皮一样以层、薄片或鳞片剥落。

脱落 树皮片状脱落的过程。

（花粉粒或孢子）外壁 花粉粒壁的最外层，由基层、柱状层、覆盖层和纹饰组成。 [212]

外果皮 （见"外果皮"）。

外皮层的 与外皮层有关的。

外皮层 根的皮层细胞的最外层。 [148]

外源的 植物体外起源的。

外来的 非本地的，原产外国，尤其是热带国家。

explicative (see **replicate**).

explosive mechanism A form of seed dispersal in which the ripe fruit suddenly opens, scattering its seeds away from the parent plant. Examples include *Impatiens* (the 5 valves of the capsule roll up), *Ulex* and other members of the Leguminosae (the 2 valves of the capsule separate and twist), and *Ecballium* (seeds ejected in a watery fluid through a hole at the end of the capsule).

exserted Protruding beyond the surrounding parts.

exstipulate Without stipules.

extine (see **exine**).

extrafloral Outside the flower, sometimes used to describe nectaries forming a ring round the reproductive organs, in other cases used to describe nectaries situated on vegetative parts of the plant.

extranuptial nectary A nectary situated away from the reproductive organs of the flower.

extrastaminal Situated outside the whorl of stamens.

extrorse With the anthers facing and opening outwards, away from the centre of the flower. **[218]**

eye One of the pits on a potato tuber containing a bud **[152]**; one of the three roundish areas at the base of a coconut. Two of these represent aborted carpels, the third (larger and softer than the others) indicates the place from which the shoot will emerge after germination. **[260]**

f. (fide) According to.

f., fil. (filius) Son, as in e.g. Linnaeus fil. or L.f., son of Linnaeus, a means of distinguishing the author of a plant name.

f. (forma) Form.

F₁ The first filial generation, i.e. the hybrids

张开的，无皱的（见"折叠的"）。

爆裂机制 一种成熟果实突然裂开，从母株上散出种子的种子散布方式。例子包括凤仙花属植物（蒴果的5枚果瓣卷起），荆豆属和豆科其他植物（荚果的2枚果瓣裂开并扭曲），喷瓜属植物（从蒴果顶端的孔以水流形式喷出种子）。

伸出的，突出的 从环绕部分以上突出。

无托叶的 没有托叶的。

（花粉粒或孢子）**外壁**（见"外壁"）。

在花外的 花的外面，有时用来描述生殖器官周围形成一圈的蜜腺，在其他情况下，用来描述位于植物营养部分上的蜜腺。

花外蜜腺 位于离开花生殖器官的蜜腺。

在雄蕊外的 位于雄蕊轮外面的。

（花药）**外向的** 具有离开花中心、向外开裂的花药。[218]

芽眼 马铃薯块茎上含有芽的凹陷之一[152]；椰子果实基部的三个圆形区域之一，其中2个代表不育心皮，第3个大而柔软，表示萌发后芽露出的地方。[260]

根据 根据。

儿子 如Linnaeus fil.或L.f.，林奈的儿子，区别植物名称作者的一种手段。

变型 变型。

第一子代，子一代 第一子代，即通常为商

produced from the first crossing of two parent plants, usually for commercial purposes.

F₂ The second filial generation, i.e. the hybrids produced by allowing certain plants of the F₁ generation to interbreed.

falcate Sickle-shaped. **[190]**

fall One of the outer perianth segments in flowers of the genus *Iris*, usually narrow at the base but expanding into a broad, pendulous blade. **[238]**

false fruit (see **pseudocarp**).

false indusium A protective covering of the sporangia formed by the reflexed margin of the frond, as in the genus *Adiantum*. **[279]**

false septum (see **replum**).

false whorl (see **verticillaster**).

family A group of genera resembling one another in general appearance and technical characters. The names of families normally end in '-aceae', but eight long-established family names not ending in '-aceae' are still in use. These are given here with their modern equivalents: Compositae (Asteraceae), Cruciferae (Brassicaceae), Gramineae (Poaceae), Guttiferae (Clusiaceae), Labiatae (Lamiaceae), Leguminosae (Fabaceae), Palmae (Arecaceae), and Umbelliferae (Apiaceae).

fan-shaped (see **flabellate**).

farina The powdery or flour-like covering found on some leaves, stems or floral parts.

farinaceous Having a powdery appearance.

farinose Covered with a fine powder. **[199]**

fasciation The abnormally broad and flattened growth of a stem caused by meristem damage, bacterial infection or mutation so that it resembles sereval stems placed side by side. The condition, which also affects the inflorescence, occurs commonly in *Taraxacum* (dandelion) and *Plantago*

业目的，由两个亲本第一次杂交产生的杂种。

第二子代，子二代 第二子代，即允许F1代的某些植物进行混交而产生的杂种。

镰刀状的 镰刀形的。[190]

瀑瓣 鸢尾属植物花的外轮花被裂片之一，基部通常狭窄但向上逐渐变宽形成宽阔而下垂的瓣片。[238]

假果 （见"假果"）。

假囊群盖 一类由叶片边缘反折形成的孢子囊的保护性覆盖物，如铁线蕨属植物。[279]

假隔膜 （见"假隔膜"）。

假轮生 （见"假轮生"）。

科 一群基本外貌和鉴别特征上彼此相似的属。科的名称通常以"-aceae"结尾，但是8个长期建立起来的、不以"-aceae"结尾的科名仍然在使用。这里列出它们及其对等名称，菊科Compositae（Asteraceae）、十字花科Cruciferae（Brassicaceae）、禾本科Graminneae（Poaceae）、藤黄科Guttiferae（Clusiaceae）、唇形科Labiatae（Lamicaeae）、豆科Leguminosae（Fabaceae）、棕榈科Palmae（Arecaceae）和伞形科Umbelliferae（Apiaceae）。

扇形的 （见"扇形的"）。

粉末 在某些叶、茎和花部分上发现的粉末状覆盖物。

粉状的 具粉末状外表的。

含粉的，具粉的 覆盖有细粉末的。[199]

扁化现象 因分生组织损伤、细菌感染或突变引起的茎不正常加宽变扁生长，以致于像若干条茎并肩排列在一起。这种情况也影响到花序，经常发生在蒲公英属植物（蒲公英）和车前草属植物（车前草）上。鸡冠花因突变而长成的奇怪花序。[183]

(plantain). The pot-plant *Celosia cristata* (Cockscomb) is grown for its curious infloreseence induced by mutation. [183]

fascicle A close cluster or bundle. [217, 269]

fascicled (see **fasciculate**).

fascicular cambium The layer of meristematic cells lying between xylem and phloem in a vascular bundle. [171]

fasciculate (fascicled) Arranged in clusters or bundles. [186]

fastigiate With the branches more or less erect and appressed, as in *Populus nigra* 'Italica' (Lombardy Poplar). [164]

feathered With feather-like markings, as those on the perianth segments of some species of *Crocus*; formed like a feather, as the bract in *Euphorbia helioscopia* (Sun Spurge). [207]

felted Densely matted, with intertwined hairs.

female flower Having a functional gynoecium, but only rudimentary or sterile stamens. [207, 227, 263]

fenestrate, fenestrated Having a fenestration. [181, 229]

fenestration (window, window pane) A translucent area in a flower or leaf. In the corolla tube of some gentians and in the lip of the orchid *Cypripedium* it is probably a guide to pollinating insects, while in the domed top of the leaf of *Darlingtonia*, a carnivorous plant, it may assist the plant to catch its insect prey. [181, 229]

fern A member of the Filicopsida, a class of flowerless plants with usually large leaves (megaphylls), often called fronds. [275—280]

fern ally A member of the Psilotopsida (fork ferns), Lycopsida (club mosses and quill worts) or Equisetopsida (horsetails), classes of flowerless plants with usually small leaves (microphylls). [280—284]

ferruginous Rust-coloured.

簇，束 紧密的簇或束。[217, 269]

簇生的，束生的 （见"丛生的，束状的"）。

束中形成层 位于维管束内木质部和韧皮部之间的分生细胞层。[171]

簇生的，束生的 成簇或成束排列的。[186]

帚状的 具多或少直立而紧靠的枝条的，像黑杨的株形。[164]

羽状的 具羽毛状斑纹的，像番红花属某些植物花被裂片那样；形状像羽毛的，像泽漆雄花的苞片。[207]

具绒毛的，毡状的 密集缠绕的，具缠绕毛的。

雌花 具有功能性雌蕊群，但仅有未完全发育的或不育的雄蕊的花。[207, 227, 263]

具窗孔的 具有窗孔的。[181, 229]

窗孔，膜孔 花或叶上的半透明区域。在某些龙胆类植物的花冠筒和兰科植物花的唇瓣上，它可能是某些传粉昆虫的向导，另外，在眼镜蛇草（一种肉食植物）叶的半球形顶端，它可以协助植物捕捉受骗的昆虫。[181, 229]

蕨类 通常具有大型叶的无花植物纲，即蕨纲的植物。[275—280]

拟蕨类 通常具有小型叶的无花植物纲，即松叶蕨纲、石松纲、木贼纲的植物。[280—284]

锈色的 锈色的。

fertile Able to reproduce sexually. **[207, 217]**

fertilisation The fusion of a male and female reproductive cell resulting in a zygote. **[213]**

festucoid Resembling the grass *Festuca* (fescue).

fetid (see **foetid**).

Fibonacci series A sequence of numbers named after the Italian mathematician Leonardo Fibonacci (c.1175-c.1250) in which each number is the sum of the preceding two numbers. It is relevant to botany in certain spiral arrangements, e.g. alternate leaves on a stem, florets in the heads of flowers in the Compositae, scales in the cones of the Gymnospermae, and the individual fruits which comprise the multiple fruit in *Ananas* (pineapple). **[175]**

fibres Elongated cells, pointed at both ends, that together with the sclereids, compose the sclerenchyma.

fibril A small fibre.

fibrous Thread-like.

fibrous root A thread-like root, particularly one of the adventitious roots arising from the base of grass stems. **[145]**

fiddlehead (see **crozier**).

fil. (see **f., fil.**).

filament A fine, elongated, thread-like structure, especially the stalk of an anther. **[216, 228]**

filiform Thread-like. **[178, 190, 220]**

fimbriate, fimbriated Fringed, having the margin cut into long, slender lobes. **[193, 213, 233]**

fish-tail leaflet A leaflet shaped more or less like the tail of a fish, as in some palms, e.g. *Caryota.* **[249]**

fish-tail nectary The forked structure, covered in nectar glands, at the mouth of the pitcher-like trap formed by the leaf of *Darlingtonia.* **[181]**

可育的 能有性生殖的。[207, 217]

受精 雄性生殖细胞和雌性生殖细胞融合产生合子。[213]

羊茅类（型） 类似于羊茅属草本植物（羊茅）的。

恶臭的 （见"具恶臭的"）。

斐氏数列 一种以意大利数学家列奥纳多·斐波那契（约1175—约1250）的名字命名的数列，在该数列中，任一数字都是它前面两个数字之和。在某些螺旋状排列次序，如茎上的互生叶、菊科植物头状花序上的小花、裸子植物球果上的种鳞和菠萝聚花果上的单个果实等方面，这个序列与植物学相关。[175]

纤维细胞 两端锐尖的伸长细胞，和石细胞一起构成机械组织。

纤丝 小纤维。

具纤维的，纤维状的 绒状的。

须根 一类丝状的根，尤其是禾本科植物茎基部生出的不定根之一。[145]

拳卷叶芽 （见"拳卷幼芽"）。

儿子 （见"儿子"）。

丝状物，花丝 一类纤细、伸长的丝状结构，特别是花药的柄。[216, 228]

丝状的 线状的。[178, 190, 220]

流苏状的，具缘毛的 具流苏的，具切成长而纤细裂片的边缘的。[193, 213, 233]

鱼尾状小叶 形状或多或少像鱼尾巴的小叶，如鱼尾葵属植物的叶。[249]

鱼尾状蜜腺 眼镜蛇草叶形成捕虫器口部覆盖着蜜腺的叉状结构。[181]

fissure A usually long, narrow opening, caused by the separation of the parts of an organ.

fistulose Cylindrical and hollow, like a pipe.

fl. (flos) Flower.

flabellate, flabelliform Fan-shaped, as the leaves of *Ginkgo* and fan palms. [191, 220]

flaccid Weak, flabby, limp.

flagellum One of the whip-like hairs on an antherozoid. [272]

flexuous Wavy. [165]

floccose With tufts of woolly hairs. [201]

flora The plant population of a particular region; a book listing and describing the plants found in a given area.

floral Relating to the flower.

floral cup (see **hypanthium**).

floral diagram A stylised drawing of the cross-section of a flower, showing the number and relative position of the various parts. [230]

floral envelope The perianth of a flower.

floral formula A system for representing the structure of a flower, using the capital letters K (calyx), C (corolla), A (androecium) and G (gynoecium) in that order. The letter P (perianth) is used instead of K and C in cases where the calyx and corolla are not clearly differentiated. Each letter is followed by a figure to indicate the number of parts of which each whorl is composed. E.g. C5=5 separate petals, C(5)=5 petals joined together, A5+5=2 whorls of stamens with 5 in each whorl, A12-20=the range of variation in the number of stamens. Sometimes a curved line above two adjacent letters is used to indicate that these whorls are joined together.

flore pleno With a double flower. [210]

florescence The part of a synflorescence that includes the main axis.

裂缝 通常由器官部分的裂开引起的长而狭窄的开口。

管状的 圆柱状、空心的，象管子。

花 花。

扇形的 像银杏和扇形棕榈的叶一样扇形的。[191, 220]

软弱的 软弱的，无力的，松弛的。

鞭毛 游动精子的鞭状毛之一。[272]

波状的 波状的。[165]

有丛毛的 羊毛状的。[201]

区系，植物志 一特定地区的植物居群；列出和描述一给定地区所发现的植物的书。

花的 与花有关的。

花杯 （见"花筒"）。

花图式 一种显示花不同部分的数目和相对关系的花横切面的程式图。[230]

花被 花被。

花程式 一种用K（花萼）、C（花冠）、A（雄蕊群）和G（雌蕊群）代表花结构的系统。在花萼和花冠不明显分化的情况下，字母P（花被）被用来代替K和C。每一字母后面的数字表示构成每轮各部分的数目。如C5=5枚离生花瓣，C(5)=5枚合生花瓣，A5+5=2轮雄蕊，每轮5枚，A12-20=雄蕊数目的变异范围。有时两个相邻字母上面的弧线用来表示这些轮是结合在一起的。

具重瓣花 具重被花的。[210]

花序 包括主轴的复合花序的部分。

floret A small flower, as in the Gramineae and Compositae. **[234, 244, 263]**

floribunda A type of garden rose that bears abundant flowers in dense clusters.

floriferous Bearing numerous flowers.

floristic Relating to a flora.

floury Powdery.

flower The structure in angiosperms concerned with sexual reproduction. **[168, 169, 225—238]**

flowerless plant A cryptogam.

fl. pl. Flore pleno. **[210]**

foetid (fetid) Having a highly unpleasant smell.

foliaceous Leaf-like; leafy.

foliage The leaves of a plant, considered as a whole.

foliar Relating to leaves.

foliose Leafy.

follicle A dry fruit formed from a single carpel, containing more than one seed, and splitting open along the suture. **[68, 177]**

follicular In the form of a follicle.

food body (see **elaiosome**)

forb Any herbaceous plant other than a grass.

form, forma (f.) A rank used to indicate a minor variant of a species, subspecies, or variety, such as a randomly occurring white-flowered plant in a population with typically coloured flowers.

fornix A small, arched scale, as those in the throat of the tubular corolla of the genus *Myosotis* (Forget-me-not).

foveate Pitted.

foveolate With small pits or depressions.

fr. (fructus) Fruit.

free Having adjoining, but different, parts completely separate from each other, e.g. as stamens from petals. **[219]**

free-central placentation The arrangement in which the placentas are situated on a central

小花 像禾本科和菊科植物的小花。 [234, 244, 263]

多花玫瑰 一种生出密集成簇的多数花朵的玫瑰园艺品种。

多花的 生有多数花的。

植物的，植物种类的 与植物区系或植物志有关的。

粉状的 粉状的。

花 与有性生殖有关的被子植物的结构。 [168, 169, 225—238]

无花植物 隐花植物。

具重被花的 具重被花的。 [210]

恶臭的 有非常不愉快的味道。

叶状的，叶的 像叶的、叶状的。

叶 植物叶的总称。

叶的 与叶有关的。

密叶的 多叶的。

蓇葖果 一类由单一心皮形成、含一枚以上种子、沿缝线开裂的干果。 [68, 177]

蓇葖的 蓇葖果状的。

食物体 （见"油质体"）。

非禾本草本植物 非禾草类植物。

变型 用来表示种、亚种或变种的小变异的等级，如在具彩色花的居群中生出白色花的植株。

冠筒鳞片 小而拱形的鳞片，如勿忘我属植物（勿忘我）花冠筒喉部的附属物。

具孔穴的 具凹的。

具小孔穴的 具小凹穴或凹陷的。

果实 果实。

离生的 具相近但不同，彼此完全离生的部分，如像雄蕊离生于花瓣。 [219]

特立中央胎座 胎座位于从一室、不被隔膜隔开的子房基部生出的中央柱上的排列

column that arises from the base of an ovary that is unilocular and not divided by septa. **[222]**

frond The leaf of a palm **[247]**, cycad **[272]** or fern. **[200]**

frondescent Breaking into leaf.

fructiferous Bearing fruit.

fructification The process of bearing fruit; the fruit of a flowering plant; the spore-bearing structures in a cryptogam.

fruit A mature ovary with its enclosed seeds and sometimes with attached external structures. **[158, 159, 169, 181, 184, 255—265]**

fruitlet One of the individual parts that compose an aggregate fruit. **[257]**

frutescent Becoming shrubby.

fruticose Shrubby.

fruticulose Somewhat shrubby.

fugacious Soon falling off or withering.

fulvous Tawny.

funicle (funiculus) The stalk connecting an ovule to its placenta. **[223]**

funicular Relating to a funicle.

funiculus (see **funicle**).

funnelform, funnel-shaped With the limb of the corolla widening gradually from a short tube. **[211]**

furcate Forked.

furfuraceous Scaly, scurfy.

furrowed Having a channel or channels along the part concerned, often broader and deeper than grooved. **[173]**

fuscous Dark brown or grey.

fused United completely. **[158, 184]**

fusiform Spindle-shaped. **[250]**

fusion Complete union of parts.

G (in a floral formula) Gynoecium, e.g. G3 indicates a gynoecium composed of 3 carpels. A line above the 'G' indicates that the ovary is inferior; a line below the 'G'

[222]

叶状体 棕榈 [247]、铁树 [272] 或蕨类植物的叶子。

叶状柄的 突变到叶的。

产果的 生有果实的。

结实 产生果实的过程；有花植物的果实；隐花植物生有孢子的结构。

果实 具封闭的种子，有时还附着有外部结构的成熟子房。 [158, 159, 169, 181, 184, 255—265]

小果 组成聚合果的单个部分之一。 [257]

近灌木状 变为灌木状的。

灌木状的 灌木状的。

小灌木状的 稍灌木状的。

先落的 不久即脱落或凋谢。

黄褐色的 黄褐色的。

珠柄 连接胚珠到胎座的柄。 [223]

珠柄的 与珠柄有关的。

珠柄 （见"珠柄"）。

漏斗状的 具有从短花冠筒逐渐变宽的花冠檐部的。 [211]

分叉的 叉状的。

软鳞片状的，具软鳞片的 鳞片状的，糠秕状的。

具沟的 沿相关部分具沟的，通常宽于并深于沟（groove）。 [173]

暗褐色的 暗褐色或灰色的。

融合的 完全结合的。 [158, 184]

纺锤状的，棱状的 纺锤形的。 [250]

融合 各部分完全融合。

雌蕊群（花程式） Gynoecium的缩写，如G3表示雌蕊群由3枚心皮组成。G上的横线表示子房是下位的；G下的横线表示子房是上位的。

indicates that the ovary is superior.

g. (gen.) Genus.

galbulus The fruit of *Juniperus* (juniper), a modified cone that becomes fleshy and berry-1ike as it matures. **[270]**

galea A hood or helmet-shaped structure formed by the perianth segments of certain flowers. **[208, 211]**

galeate With a galea. **[129]**

gall Abnormal growth of plant tissue in response to an attack by insects, fungi, bacteria, mites etc., the characteristic form of the gall often revealing the cause of the abnormality. **[182, 263]**

gamete One of the male or female sex cells (usually haploid) that unite at fertilisation to form a zygote.

gametophyte The sexual stage in the life cycle of a plant when the chromosomes in each cell are reduced to half the usual number, typically diploid reduced to haploid. **[275]**

gamodeme A deme forming a relatively isolated community.

gamopetalous (sympetalous) Having petals that are united at least at the base, as *Primula veris*. **[229]**

gamophyllous Having united petals and sepals.

gamosepalous (synsepalous) Having sepals that are united at least at the base, as *Primula veris*. **[229]**

geitonogamy A form of allogamy, in which the ovules of a flower are fertilised by pollen from another flower on the same plant.

gelatinous Jelly-like.

geminate (paired) Arranged in pairs. **[170, 184, 226, 267, 271]**

gemma An adventitious bud arising on a fern frond that can develop into a plantlet. **[282]**

gemmiferous, gemmiparous Bearing gemmae.

gen.(g.) Genus.

属 属。

柏树球果 刺柏属植物（刺柏）的果实，一类成熟时肉质、浆果状的变态球果。[270]

盔瓣 某些花的花被裂片形成的兜帽状或头盔状结构。[208, 211]

盔状的 具盔的。

虫瘿 植物组织受到昆虫、真菌、细菌和螨虫等侵染而引起的不正常生长物，虫瘿的特征性形状通常揭示出这种不正常发育的原因。[182, 263]

配子 受精时结合形成合子的雄性或雌性细胞（通常单倍体）之一。

配子体 植物生活史中的有性阶段，每个细胞的染色体数目减少到平常的一半，典型地由二倍体减少到单倍体。[275]

隔离同类群 形成相对隔离群落的同类群。

合瓣的 具有至少基部结合花瓣的，如黄花九轮草。[229]

合被的 具有结合的花瓣和萼片。

合萼的 具有至少基部结合萼片的，如黄花九轮草。[229]

同株异花受精 异花受精的一种形式，即一朵花的胚珠被来自同一植株上另一花的花粉受精。

胶质的 胶状的。

成对的 成对排列的。[170, 184, 226, 267, 271]

胞芽 一类从蕨类植物叶状体上生出的、能够发育成小植株的不定芽。[282]

具胞芽的 生有胞芽的。

属 属。

gene One of the units of heredity occupying a fixed position (locus) on a chromosome, that either by itself or in combination with other genes is responsible for a particular characteristic, e.g. height, flower colour, etc.

基因 遗传单位之一，占据染色体上一个固定的位置，通过自身或与其他基因结合控制一对特定性状，如高度、花色等。

generative cell The smaller of the two cells into which the nucleus of the pollen grain divides while still in the pollen sac. This cell subsequently divides, sometimes before pollination, and gives rise to two male nuclei (sperm cells). **[213]**

生殖细胞 花粉粒的核分裂形成的两个细胞中的较小者，这个细胞继续分裂，有时在传粉前，产生两个雄核（精子细胞）。[213]

genet One or more individuals produced by asexual reproduction from a single zygote.

无性株 从一个合子通过无性繁殖产生的一个或多个个体。

genetic Relating to genes.

遗传的 与基因有关的。

genetics The study of variation and heredity.

遗传学 变异和遗传的研究。

geniculate Abruptly bent or "knee-like". **[220, 241]**

膝曲的 急弯曲的，或膝状的。[220, 241]

genus (plural **genera**) A botanical rank, comprising one or more similar species. The names of genera are written with a capital initial letter.

属（复数为genera） 一植物学等级，包括一个或多个相似的种。属名用大写首字母印刷。

geocarpic With fruits ripening below ground, as *Arachis hypogaea* (Groundnut). **[174]**

地下结果 具有地下成熟果实，如落花生。[174]

geocarpy The ripening of fruits below ground from flowers borne above the ground, the young fruits being pushed into the soil by the curving action of the stalk. **[174]**

地下结果性 从地面之上着生的花，通过子房柄的弯曲活动将幼果推向土壤，在地下成熟。[174]

geophyte An herbaceous plant that perennates by means of underground buds, e.g. bulbs, corms, etc.

地下芽植物 一类通过地下芽如鳞茎、球茎等多年生长的草本植物。

geotropic (gravitropic) Responding to gravity, either positively geotropic as roots that grow downwards, or negatively geotropic as stems that grow upwards. **[174]**

向地性的 对重力的响应，既有像根向下生长的正向地性，像茎向上生长的负向地性。[174]

geotropism (gravitropism) The growth movement of plants in response to gravity. **[174]**

向地性 植物对重力响应的生长运动。[174]

germ cell A cell specialised for reproductive purposes, which gives rise to male or female gametes.

生殖细胞 为生殖目的而特化的产生雌雄配子的细胞。

germination The development of a spore into

萌发，发芽 孢子发育成原叶体或种子发育

a prothallus or a seed into a seedling. **[157, 213]**

gibbose, gibbous With a pouch-like swelling. **[211]**

Ginkgoopsida The class that includes *Ginkgo*. **[271]** (see **Gymnospermae**)

girdle scar The scar left by the terminal bud of the previous year. **[170]**

glabrate Almost hairless.

glabrescent Becoming hairless.

glabrous Without hairs.

gland A organ producing a secretion. **[177, 207]**

glandiferous Bearing glands.

glandular Possessing glands.

glandular hair A hair with a gland at its apex. **[169, 201]**

glaucescent Becoming glaucous, but sometimes incorrectly used to mean slightly glaucous.

glaucous With a waxy, greyish blue bloom.

globose, globular Spherical or globe-shaped, as the flower of *Trollius europaeus* (Globe Flower). **[210, 270]**

glochid (see **glochidium**).

glochidiate Bearing barbed hairs or bristles.

glochidium (glochid) A barbed hair or bristle. **[240, 281]**

glomerule A compact cluster of cells or spores; a condensed cyme of almost sessile flowers.

glossy (see **lustrous**).

glume One of the pair of bracts at the base of a spikelet in the Gramineae **[243]**; the single bract subtending the flower in the Cyperaceae. **[246]**

glutinous Sticky.

Gnetopsida The class containing the families Ephedraceae, Gnetaceae, and Welwitschiaceae. **[196]** (see **Gymnospermae**)

graft A portion of a plant inserted into and

成幼苗。[157, 213]

具囊状膨大的 具袋状膨胀的。[211]

银杏纲 包括银杏的纲。[271]（见"裸子植物"）

环痕 去年顶芽留下的痕迹。[170]

平滑的，光滑的 几乎无毛的。

近无毛的 变无毛的。

平滑的，无毛的 光滑无毛的。

腺（体） 产生分泌物的器官。[177, 207]

具腺的 生有腺体的。

具腺体的 具有腺体的。

腺毛 顶端具有腺体的毛。[169, 201]

带白霜的 变苍白色的，但有时错误地用来表示浅苍白色。

具白霜的 具蜡质、灰蓝色粉末的。

球形的 圆形或球状的，像球兰的花。[210, 270]

钩毛，倒刺毛。（见"倒刺毛"）。

具倒钩毛的 生有钩状毛或刚毛的。

钩毛 钩状毛或刚毛。[240, 281]

细胞群，孢子群，团伞花序 细胞或孢子构成的密集簇，或近无柄花构成的紧缩聚伞花序。

有光泽的 （见"有光泽的"）。

颖片 禾本科植物小穗基部的一对苞片之一 [243]；莎草科植物包在花下的单一苞片。[246]

胶黏的 黏的。

买麻藤纲 包括麻黄科、买麻藤科和百岁兰科的纲（见"裸子植物"）。

嫁接 植物体的一部分插进并连接到另一

uniting with a larger part of another plant, as a scion into a stock.

植物体的更大部分上，像插穗接到砧木上。

graft chimaera, graft hybrid A plant that has originated by grafting rather than by sexual reproduction, e.g. +*Laburnocytisus adamii*, derived from *Laburnum anagyroides* and *Cytisus Purpureus*.

嫁接嵌合体，嫁接杂种 一类来源于嫁接、而不是有性生殖的植物，如毒雀花，来源于金链花和紫金雀花的嫁接。

grain (see **caryopsis**).

谷粒，颖果 （见"颖果"）。

graminaceous, gramineous Relating to the Gramineae.

禾本科的 与禾本科有关的。

graminoid Grass-like.

禾草状的 禾草状的。

graniferous Bearing grain, or seed resembling grain.

结谷粒的 生有谷粒，或类似谷粒状种子的。

granular, granulose Having a slightly rough surface.

具颗粒的 具有稍粗糙表面的。

grex (plural **greges** or **grexes**) A collective name covering all the progeny of two parent plants, and now used only for orchid hybrids, e.g. *Paphiopedilum* Maudiae. It is printed in Roman type with a capital initial letter.

杂种后代（复数为greges或grexes） 涵盖两亲本植株所有后代的集合名称，现在仅指兰花杂交种，如魔帝兜兰。用具有大写首字母的罗马体印刷。

grooved Having a channel or channels along the part concerned, often narrower and shallower than furrowed. [173]

具沟的 具沟的或沿相关部分具沟的，通常窄于并浅于犁沟（furrow）。 [173]

ground cover The lowest layer of vegetation in a wood or forest, consisting of herbaceous plants sometimes specially planted.

地被植物，地被物 森林植被的最下层，包括草本植物，有时特别栽植的。

ground tissue Tissue other than vascular tissue, e.g. pith, cortex, etc.

基本组织 不同于维管组织的组织，如髓、皮层等。

growth ring (see **annual ring**).

生长轮 （见"年轮"）。

grp. Group.

组 组。

guard cell One of the pair of specialised cells of a stoma that control the size of the aperture and regulate the flow of gases in and out of the plant. [196, 268]

保卫细胞 气孔的一对控制气孔大小、调节气体出入植物体的特化细胞之一。 [196, 268]

guttation The secretion of droplets of water from a plant, typically from hydathodes at the tips or on the margins of leaves.

吐水作用 从植物体，典型的是从叶尖或叶边缘的排水器，分泌出小水滴。

Gymnospermae, Gymnospermophyta The gymnosperms, a group comprising Cycadopsida (cycads), Ginkgoopsida,

裸子植物，裸子植物门 裸子植物，一类胚珠裸露，也就是不包被在子房内的植物，包括苏铁纲、银杏纲、买麻藤纲和

Gnetopsida, and Pinopsida (conifers), plants whose ovules are naked, i.e.not enclosed in an ovary. **[267—273]**

gynaecium (see **gynoecium**).

gynandrous Having the stamens and style(s) united, as members of the Orchidaceae.

gynobasic With the style arising from below the ovary and between the carpels, as in many genera of the Boraginaceae and Labiatae. **[220]**

gynodioecious A species in which individual plants bear only female flowers or only bisexual flowers.

gynoecium (gynaecium) The female sex organs (carpels) collectively. In some species, e.g. *Vicia faba* (Broad Bean) the gynoecium consists only of a single carpel, but in most plants the gynoecium is composed of several carpels. **[213, 219]** (see **apocarpous** and **syncarpous**)

gynomonoecious Having female and bisexual flowers on the same plant.

gynophore The stalk bearing a carpel or gynoecium.

gynostegium The staminal crown in the flowers of some members of the Asclepiadaceae. **[233]**

gynostemium The column formed by the androecium and gynoecium combined, as in the flowers of some members of the Aristolochiaceae. **[225]**

habit The general appearance of a plant.

habitat The location in which a plant normally grows, determined by type of soil, amount of water, temperature, and other environmental factors.

haft The lower, usually narrower, part of the falls or standards in flowers of the genus *Iris*. **[238]**

half-epigynous, half-inferior (see **semi-**

球果纲。 [267—273]

雌蕊群 （见 "雌蕊群" ）。

雌雄蕊合体的 雄蕊和花柱结合的，如兰科植物。

花柱茎生的 花柱自子房下面、心皮之间生出，像紫草科和唇形科的许多属植物。[220]

雌花两性花异株的 植株上仅有雌花或仅有两性花。

雌蕊群 总体的雌性器官（心皮），在某些植物如蚕豆，雌蕊群仅有一枚心皮，但绝大多数植物，雌蕊群有数枚心皮组成。 [213, 219]（见 "离生心皮的" 和 "合生心皮的" ）

雌花两性花同株的 同一植株上具有雌花和两性花。

雌蕊柄 着生一枚心皮或雌蕊群的柄。

合蕊冠 萝藦科某些植物花的雄蕊的帽状物。[233]

合蕊柱 雄蕊群和雌蕊群结合形成的圆柱状结构，像马兜铃科某些植物的花。[225]

习性，体型 植物的总体外貌。

生境 由土壤类型、水分多少、温度和其他环境因子决定的、植物正常生长的场所。

瓣柄 鸢尾属*Iris*植物花的外花被片下部通常狭窄的部分。[238]

半上位的，半下位的（见 "半下位的" ）。

inferior).

half-parasite (see **hemi-parasite**).

halophyte A plant that is adapted to grow in saline soils.

halophytic Growing in saline soils.

hamate Hooked.

hamulate Bearing small hooks.

hapaxanthic (see **monocarpic**).

haplochlamydeous (see **monochlamydeous**).

haploid Having half the usual number of chromosomes in each cell, typically a single set.

haplostele A type of protostele consisting of a central core of xylem surrounded by a ring of phloem. [280]

haplostemonous With a single series of stamens in one whorl.

haptotropic (see **thigmotropic**).

hardiness The ability of a plant to withstand unfavourable conditions, especially cold.

hardwood Wood obtained from broad-leaved, dicotyledonous trees.

hastate Spearhead-shaped, with basal lobes directed outwards. [191, 194]

hastula A flap of tissue, borne at the junction of the petiole and the lamina in some palm leaves. [248]

haulm The stem of various herbaceous plants, e.g. peas, beans, potatoes, hops, and grasses.

haustorium An outgrowth from a parasitic plant that enables it to absorb nutrients from its host. [147]

head A short, dense spike of flowers; the capitulum in the Compositae. [205, 234, 235]

heartwood (duramen) The inner, older layers of wood in the trunk or branch of a tree or in the stems of a shrub, usually denser and darker than the surrounding sapwood, and no longer able to conduct sap. [171]

半寄生（见半寄生）。

盐土植物 适于生长在盐性土壤上的植物。

盐生的 生长在盐性土壤上的。

具钩的 钩状的。

具小钩的 具小钩的。

一次开花的（见"一次结实的"）。

单轮花被的（见"单被花的"）。

单倍体 每一细胞具有平常数目一半的染色体，一般一套染色体。

单中柱 一种包括中心木质部、其外围绕一轮韧皮部的原生中柱。[280]

具单轮雄蕊的 每轮具单一系列的雄蕊。

向触的 （见"向触的"）。

耐性 植物承受不适宜条件的能力，尤其是寒冷。

阔叶材 阔叶双子叶植物树木的木材。

戟形的 矛头状的，具有向外伸展的基部裂片。[191, 194]

叶舌 生于某些棕榈类植物叶柄与叶片连接处的扁平组织。[248]

秸，茎秆 不同草本植物，如豌豆、大豆、马铃薯、葎草和禾草类植物的茎。

吸器 寄生植物生出的能使其从寄主体内吸取营养物质的生长物。[147]

头状花序 短而密集的穗状花序；菊科植物的头状花序。[205, 234, 235]

心材 树干或树木分枝或灌木茎内的内侧老木质层，通常较周围的液材更密集，颜色更暗，不久即不能输导液体。[171]

heel The small piece of tissue that is pulled away from the parent stem when a cutting is taken from a plant.

heliciform Coiled like a snail shell.

helicoid cyme (see **bostryx**).

heliophilic Sun-loving or light-loving.

heliophobic Shade-loving.

heliophyte A plant adapted to living in high light intensities.

heliotropic Turning towards the sunlight (positively heliotropic) or away from it (negatively heliotropic). **[174]**

heliotropism The growth response of a plant to the stimulus of sunlight. **[174]**

helophyte A marsh plant, with resting buds below the surface of the marsh.

hemicryptophyte A plant with resting buds at or near the level of the soil. **[166]**

hemi-parasite (half-parasite, partial parasite, semi-parasite) A plant that derives some of its nourishment from a host plant. **[148]**

hemi-parasitic Partially parasitic, as members of the Loranthaceae, e.g. *Viscum album* (Mistletoe), and some genera of subfamily Rhinanthoideae in the Scrophulariaceae. **[148]**

hemitropous (see **amphitropous**).

heptaploid (7n) Having seven sets of chromosomes in each cell.

herb A non-woody plant, or one that is woody only at the base.

herbaceous Composed of soft, non-woody tissue. **[169]**

herbarium A collection of dried plants or parts of plants, usually mounted on sheets of thick paper of a uniform size, and kept for purposes of reference or research; the building in which such a collection is housed.

herbarium label A label affixed to an

插条 当从植物上做切割时，从亲本茎上切下的一小块组织。

螺状的 盘绕成蜗牛壳状的。

螺形聚伞花序 （见"螺状聚伞花序"）。

适阳的，嗜光的 喜阳的，好光的。

避阳的，嫌阳的 嫌阳的，避阳的。

阳生植物 一类适于生长在高光强度下的植物。

向光性的，趋光性的 转向阳光（正向阳性的），或远离阳光（负向阳性的）。[174]

向日性，趋日性 植物对阳光刺激的生长响应。[174]

沼泽植物 休眠芽位于沼泽面一下的植物。

地面芽植物 休眠芽在或接近土壤表面的植物。[166]

半寄生 从寄主植物吸取某些营养的植物。[148]

半寄生的 部分寄生的，像桑寄生科植物，如白果槲寄生，和玄参科鼻花亚科某些属植物。[148]

横生的 （见"横生的"）。

七倍体（7n） 每个细胞具有七套染色体。

草本植物 一类不木质化或仅基部木质化的植物。

草本的 由柔软而不木质化的组织组成的。[169]

腊叶标本，植物标本馆 一类干植物或干植物部分的收藏品，通常装订到同样大小的硬纸板上，长久保存，用于参考和研究；贮藏这些收藏品的建筑物。

标本记录签 粘贴在标本上显示仅从标本研

herbarium sheet giving information about the plant which cannot be obtained from study of the specimen alone, e.g. the collector, place and date of collection, dimensions (if a large plant), etc., also details which may change in the course of time, e.g. colour of flowers.

hermaphrodite (see **bisexual**).

hesperidium A berry in which the fleshy part is divided into segments and the outer skin is a tough, leathery rind, e.g. *Citrus* (orange, lemon etc.). **[260]**

heterobrochate Of pollen grains, having reticulate sculpturing, but with the meshes of the network differing in size.

heterochlamydeous Having a perianth that is clearly divided into calyx and corolla.

heterogamous With each flower-head composed of two or more kinds of flowers, as in many species of Compositae. **[234]**

heterogeneous Composed of dissimilar parts.

heteromorphic, heteromorphous Having two or more different forms.

heterophyllous Having more than one kind of leaf on the same plant. **[175]**

heterophylly The condition of being heterophyllous. **[175]**

heterosis (see **hybrid vigour**).

heterosporous Producing different kinds of spores, typically microspores and megaspores.

heterostylous Having variation in the length of the style (and stamens) in different flowers of the same species. **[229]**

heterostyly Variation in the length of the style (and stamens) in different flowers of the sarne species, as in some genera of the Primulaceae and Lythraceae. **[229]**

hexamerous Having the parts of the flower in sixes.

hexandrous Having six stamens.

究不能获取的植物信息的标签，如采集人、采集地点、采集日期和尺寸（若为大植物）等，还有随时间延续可以改变的详细特征，如花的颜色。

雌雄同体的。（见"两性的"）。

柑果 一类肉质部分分裂为许多果瓣，外果皮粗糙、革质的浆果，如柑橘属植物（橘子、柠檬等）。[260]

异型网状的 花粉粒具网状雕纹，但网眼大小不等。

异形被花的 具有明显分化为花萼和花冠的花被的。

具异形花的 每一头状花序均由两种或更多种类型的花构成的，像菊科许多种植物。[234]

杂的，异种的 由不同部分组成的。

异形的 具有两种或更多不同形式的。

具异形叶的 同一植株上具有一种以上形态的叶。[175]

异形叶性 具异形叶的状态。[175]

杂种优势 （见"杂种优势"）。

具异形孢子的 产生不同类型的孢子，通常为小孢子和大孢子的。

花柱异长的 同种植物不同花的花柱（和雄蕊）长度存在变异的。[229]

花柱异长 同种植物不同花的花柱（和雄蕊）长度的变异，如同报春花科和千层菜科的某些层。[229]

六基数的 花各部分数目均为六的。

六枚雄蕊的 具有六枚雄蕊的。

hexaploid (6n) Having six sets of chromosomes in each cell.

hexarch root A root with six protoxylem strands in the stele.

hibernaculum A winter bud, formed when the plant dies down, and from which it re generates.

hibernal Occurring in winter.

hilum The scar left on a seed where it was previously attached to the funicle. [157]

hinge cell (see **bulliform cell**).

hip The false fruit or pseudocarp in the genus *Rosa* (rose), developed from the fleshy, hollow hypanthium and containing achenes. [261]

hippocrepiform Horseshoe-shaped.

hirsute Covered in rough, coarse hairs. [200]

hirsutulous Slightly hirsute. [201]

hirtellous Minutely hirsute. [201]

hispid Having small, stiff, bristly hairs. [200]

hispidulous Having small, stiff, bristly hairs.

histogen One of the three layers (dermatogen, periblem, and plerome) considered to be present in an apical meristem, more recently challenged by the concept of tunica and corpus.

hoary Covered with small, whitish hairs, giving the surface a frosted appearance.

homochlamydeous (homoiochlamydeous) Having a perianth composed of similar segments, and therefore not clearly divided into calyx and corolla. [168] (see **tepal**)

homogamous With only one kind of flower; with anthers and stigmas maturing simultaneously.

homogeneous Of uniform structure, com posed of similar or identical parts.

homoiochlamydeous (see **homochlamydeous**).

homologous Similar in structure and origin.

homosporous Producing only one kind of spore.

六倍体（6n）的 每个细胞中具有六套染色体的。

六原型根 中柱内具有六条原生木质部的根。

越冬芽 植物在冬季临近死亡时形成的、来年再萌发的芽。

冬出的 在冬季显现的。

种脐 在种子上留下的以前着生到珠柄的痕迹。[157]

铰合细胞 （见"泡状细胞"）。

蔷薇果 蔷薇属植物（玫瑰）由肉质、中空的花托筒发育而来的假果，内含瘦果。[261]

马蹄形的 马蹄形的。

具长硬毛的 被粗糙毛的。[200]

具微糙硬毛的 稍具长硬毛的。[201]

具微糙硬毛的 具小长硬毛的。[201]

具糙硬毛的 具有坚硬、刚毛状毛的。[200]

具短硬毛的 具有小、坚硬、刚毛状毛的。

组织原 被认为是展示顶端分生组织的三层结构（表皮原、皮层原和中柱原）之一，近来更多地受到原套—原体学说的挑战。

灰白毛的 覆盖着小而灰白色毛，表面呈现为磨砂外观。

同形花被的 具由相同花被裂片构成的花被，因此不能明显地分为花萼与花冠。[168]（见"花被片"）

具同形花的，雌雄同熟的 仅具一种花的；具同时成熟的花药和柱头的。

同形的，同型的 具相同部分组成的一样结构的。

同形花被的 （见"同被花的"）。

同源的 结构和起源相同的。

具同形孢子的 只产生一种孢子的。

homostylous Having styles of the same length.

花柱等长的 具有相同长度的花柱。

homostyly The state of being homostylous.

花柱等长 具花柱等长的状态。

honey gland (see **nectary**).

蜜腺 （见"蜜腺"）。

honey guide (see **nectar guide**).

蜜指示 （见"蜜指示"）。

honey leaf A nectary, as in many members of the Ranunculaceae. **[226]**

蜜叶 一种蜜腺像毛茛科许多植物。 [226]

hood The rounded lid that forms a canopy over the mouth of the pitcher-shaped leaves in e.g. *Darlingtonia* and *Sarracenia*. **[181]**

兜状瓣 如眼镜蛇草属和瓶子草属的袋状叶口部的圆形顶盖。 [181]

horizontal With the branches growing at right-angles to the trunk. **[163, 164]**

水平的 具与树干呈直角生长的分枝。 [163, 164]

hormone A substance formed in a plant which has a specific effect on its growth or development. **[173]**

荷尔蒙，激素 植物体内形成的、对其生长或发育具有特殊作用的一类物质。 [173]

hose-in-hose The unusual arrangement of flowers in some forms of *Primula vulgaris* (Primrose) and *P.veris* (Cowslip), in which the flowers are in pairs, one growing from the centre of the other. **[183]**

套冠，重叠花 欧洲报春和黄花九轮草的某些变型花的异常排列方式，花成对生长，一朵花生到另一朵花的中间。 [183]

host A plant from which a parasite obtains nutrients. **[147, 148]**

寄主 寄生生物从其获取营养的植物。 [147, 148]

husk The external, membranous covering of certain seeds; one of the bracts that surround and protect the female inflorescence of *Zea mays* (Maize, Sweet Corn). **[244]**

外壳，外果壳 某些种子外面的膜质覆盖物；如围玉米雌花序周围的起保护作用的苞片之一。 [244]

hyaline Thin, colourless, and translucent.

透明的 薄、无色而透明的。

hybrid A plant resulting from a cross between two or more plants, genetically unlike and belonging to different taxa, e.g. *Geum* x *intermedium*, a cross between two species in the same genus (*G.rivale* and *G.urbanum*), or x *Pyronia veitchii*, a cross between two species in difierent genera (*Cydonia oblonga* and *Pyrus communis*). **[210]**

杂种 遗传上不相同、属于不同类群的两种或多种植物间杂交产生的植物，如过渡路边青*Geum×intermedium*是同属两种植物紫萼路边青*G. rivale*和菩提香*G. urbanum*杂交的种；楒梓梨（*×Pyronia veitchii*）是不同属的两种植物楒梓（*Cydonia oblonga*）和西洋梨（*Pyrus communis*）杂交的种。 [210]

hybrid vigour An increase in desirable characteristics, e.g. growth rate, yield, etc., exhibited by hybrids in comparison with their parents.

杂种优势 杂种与亲本相比展现出来的在生长速度、产量等理想性状上的增长。

hydathode A pore or gland that exudes water.

排水器 排除水分的孔或腺体。

hydrochoric, hydrochorous Having seeds

水播的 具水播种子的。

that are dispersed by water.

hydrochory Dispersal of seeds by water.

hydromorphic Exhibiting hydromorphy.

hydromorphy The specialised structure present in the submerged stems and leaves of aquatic plants.

hydrophilous Depending on water to convey pollen for fertilisation.

hydrophily Pollination by water.

hydrophyte An aquatic plant, one that grows in water or needs a waterlogged habitat. **[183]**

hydrophytic Growing in a wet environment. **[175]**

hydrotropic Turning towards the source of water. **[174]**

hydrotropism The growth of a plant in response to the stimulus of water, as when the root of a plant turns towards the source of moisture. **[174]**

hygrochastic A type of plant movement resulting from the absorption of water, as capsules that open in moist air.

hygromorphic Exhibiting hygromorphy.

hygromorphy The specialised structure present in land plants growing in very damp habitats which promotes transpiration.

hygrophilous Requiring abundant moisture.

hygrophyte A land plant adapted to a perpetually damp habitat.

hygroscopic Extending or shrinking according to changes in moisture content.

hypanthial Relating to a hypanthium.

hypanthium (floral cup) A cup-shaped or tubular enlargement of the receptacle or of the bases of the floral parts. **[228, 261]**

hypochile The basal portion of the labellum in an orchid. **[253]**

hypocotyl The part of a seedling below the cotyledons which gives rise to the root. **[153, 157, 159]**

hypocrateriform (salverform, salver-shaped)

（植物或种子）水播 靠水传播种子。

水生形态的 展示水生形态的。

水生形态 存在于水生植物沉水茎叶内的特化结构。

水媒的 依靠水传播花粉进行受精的。

水媒 靠水传粉。

水生植物 一类生长于水中或需要水淹生境的植物。 [183]

水生的 在潮湿环境生长的。 [175]

向水的 转向水源的。 [174]

向水性 当植物根转向潮湿源时，为响应水刺激而进行的植物的生长。 [174]

吸湿开裂的 通过吸水导致的一种植物运动的形式，如蒴果在潮湿空气中开裂。

吸湿性的 展示吸湿性的。

吸湿性 生在非常潮湿生境的陆生植物的促进蒸腾作用的特化结构。

适湿的，嗜湿的，需要充分潮湿的。

湿生植物 适于持久的潮湿生境的植物。

吸湿的 根据潮湿度的改变伸展或收缩。

托杯的 与花托筒有关的。

花托筒，托杯 花托或花各部分基部的杯状或筒状膨大物。 [228, 261]

唇瓣基 兰花唇瓣的基部。 [253]

下胚轴 子叶以下、生出根的幼苗部分。 [153, 157, 159]

托盘状的（高脚碟状的） 具有一纤细花冠

Having a slender tube that expands abruptly into a flat or saucer-shaped limb. [211]

hypodermis The layer of cells immediately below the epidermis. [268]

hypogeal A type of germination in which the cotyledons remain underground. [157, 158]

hypogynous With the sepals, petals and stamens attached to the receptacle or axis below the ovary. [218]

hyponasty The tendency in part of a plant to grow more rapidly on the lower side so that it curves upwards.

hypopodium The lower portion of a sylleptic shoot from the adjoining stem to the first leaf or leaves.

hypostomatal, hypostomatic With stomata only on the lower surface of the leaf.

hysteranthous Describes leaves which are produced after the plant has flowered.

I.C.B.N. The International Code of Botanical Nomenclature, a set of internationally accepted rules, first published in 1952, that govern the naming of botanical taxa. These regulations and recommendations are amended when necessary at the International Botanical Congresses that are held in different countries every five or six years.

i.e. (id est) That is.

imbricate Overlapping, like fish scales or roof tiles. Some kinds of overlapping have been given separate names, e.g. contorted, quincuncial. [170, 183, 186, 189, 194]

imbricate-ascending Having the vexillum within the lateral petals in bud, as in members of subfamily Caesalpinioideae in the Leguminosae. [255]

imbricate-descending (vexillary) Having the vexillum outside the lateral petals in bud, as in most members of subfamily Papilionoideae in the Leguminosae. [255]

筒，其顶端突然扩展为一开展的、浅碟形的檐部。[211]

下皮层 紧接表皮下面的细胞层。[268]

子叶留土的 子叶留在地下的一种萌发形式。[157, 158]

（花瓣、萼片、雄蕊等）下位的 具着生于子房下花托或轴的萼片、花瓣和雄蕊的。[218]

偏下性 植物部分下侧生长更快致使向上弯曲的趋势。

柄 从着生茎到第一叶的枝条的下部。

气孔下生的 仅在叶下表面具气孔的。

花后展叶的 形容植物开花后产生叶的。

国际植物命名法规 一部1952年首次出版的、管理植物类群命名的国际承认的规则。这些规则和附则当必要时在每5年或6年一次、在不同国家召开的国际植物学大会上进行修改。

即，也就是 那就是。

覆瓦状的 像鱼鳞或屋顶瓦一样重叠覆盖的。一些类型的重叠覆盖被给予了分别的名称，如扭曲的、双盖覆瓦状的。[170, 184, 186, 189, 194]

上向覆瓦状的 在花芽内旗瓣位于两个侧生花瓣之内，像豆科苏木亚科植物。[255]

下向覆瓦状的 在花芽内旗瓣位于两个侧生花瓣之外，像豆科蝶形花亚科植物。[255]

immaculate Withoat any spots or markings.

imparipinnate (odd-pinnate) Pinnate, with a terminal leaflet. **[192]**

imperfect flower A flower in which only the androecium or the gynoecium is functional.

implexed Entangled, as the hairs on some species of *Stachys* (Labiatae).

inaperturate Without openings or pores.

incised Cut deeply and sharply into narrow, angular divisions. **[193]**

included Not projecting, contained within another organ.

incompatible Unable to produce viable offspring despite the presence of fertile gametes, e.g. pollen, although functional, may not grow down the style of a particular flower and will therefore fail to fertilse the ovule.

inconspicuous Not easily seen, blending with the surrounding parts.

incumbent Lying close along a surface; of cotyledons, having dorsal sides parallel to the radicle. **[160]**

incurved Curved inwards.

indefinite Of a large enough number to make a precise count difficult.

indehiscent Remaining closed at maturity.

indeterminate A form of inflorescence in which the outer or lower flowers open first and the stem or branch continues to grow, e.g. a spike or raceme. **[204]**

indigenous (native) Occurring naturally in the region concerned.

indumentum The covering of hairs or scales.

induplicate Folded inwards or upwards. **[248]**

indurate Hardened.

indusiate Having an indusium. **[275]**

indusium A protective covering of the sporangia formed by an outgrowth from the frond. **[275]**

ined. (ineditus) Unpublished.

无斑点的 无斑点或标记的。

奇数羽状的 羽状，具有顶生小叶的。[192]

不完全花 只有雄蕊群或雌蕊群具有功能的花。

交织的，缭绕的，卷入的，缠绕的 像水苏属（唇形科）某些植物体上的毛一样。

无萌发孔的 没有开口或细孔的。

锐裂的，具缺刻的 深入而尖锐地切成狭窄而角状裂片的。[193]

内藏的 不突出，包括在另一器官内的。

不亲和的 尽管可育配子存在也不能产生可育的后代，如花粉即便可育也不能沿特定花的花柱向下生长，因此将不能对胚珠授精。

不明显的 不易看到的，混在周围部分内。

子叶背倚的 紧倚表面的；具平行于胚根的背面的子叶的。[160]

内弯的 向内弯曲的。

无定数的，无限的 一个足够大的数字使准确计算非常困难的。

不开裂的 成熟时保持闭合状态的。

无限的 一类花序类型，外部的或下面的花首先开放，同时茎和分枝继续生长，如穗状花序或总状花序。[204]

原生的 自然生长于相关地区的。

毛被 毛被或鳞被。

内向镊合状的 向内或向上折叠的。[248]

硬化的 变硬的。

具囊群盖的 具有囊群盖的。[275]

囊群盖 通过叶片生长物形成的孢子囊的保护性覆盖物。[275]

未出版的 没有出版的。

inferior Below, as when the ovary appears embedded in the pedicel below the other floral parts. **[218, 225]**

infertile Not fertile.

inflated Enlarged, as the calyx of *Physalis alkekengi* (Chinese Lantern). **[210, 255]**

inflexed Abruptly bent inwards.

inflorescence The arrangement of flowers on the floral axis; a flower cluster. **[169, 183, 184, 204—209]**

infrafoliar Borne below the leaves. **[247]**

infrapetiolar Borne below the petiole.

infraspecific Below the rank of species.

infructescence A cluster of fruits, derived from an inflorescence. **[169]**

infructuous Not bearing fruit.

infundibular, infundibuliform Funnel-shaped. **[211]**

initial A cell in a meristem that divides into two daughter cells, one of which adds to the tissues of the plant, the other remaining in the meristem to repeat the process.

insectivorous plant (see **carnivorous plant**).

inserted Growing out from another part of the plant.

insertion The place where one plant part grows out of another.

in sicco In a dried state (as a herbarium specimen), a phrase used to notify possible differences from descriptions of the plant in its living condition.

in situ In the natural or original position. **[231]**

integument The outer covering of an ovule, which becomes the testa of the seed. **[213, 221, 271]**

intercalary growth The result of meristem activity between the apex and base of a stem or other part of a plant.

intercellular Between the cells.

interfascicular cambium The layer of cambium between the vascular bundles that

下位的 在下面，如子房嵌入花柄内，在花其他部分之下的。[218, 225]

不育的 不能生育的，不结果实的。

膨大的 增大的，如酸浆的花萼。[210, 255]

内折的 突然向内弯折的。

花序 花在花轴上的排列方式；花簇。[169, 183, 184, 204—209]

叶下的 生于叶下面的。[247]

柄下的 生于叶柄下的。

种下的 种等级之下的。

果序 从花序演变而来的果簇。[169]

不结果的 不产生果实的。

漏斗状的 漏斗状的。[211]

原始细胞 一类分生组织的细胞，分裂为两个子细胞，一个添加到植物组织中，另一个保留在分生组织中重复分裂。

食虫植物 （见"肉食植物"）。

着生的 从植物体另一部分生长出的。

着生处，着生点 一个植物体的部分从另一植物部分生出的位置。

干燥 在干燥状态（像腊叶标本），一个用于表示与生活状态下植物描述可能的不同的短语。

原地，原位 在自然或原始状态下。[231]

珠被 将来变成种皮的胚珠的外层覆盖物。[213, 223, 271]

居间生长，节间生长 茎的顶端和基部之间或植物体的其他部分的分生组织细胞活动的结果。

胞间的 细胞之间的。

束间形成层 位于维管束之间的形成层，它和束中形成层联合形成根或茎内的分生

joins with the fascicular cambium to form a cylinder of meristematic cells in stems and roots. **[171]**

interfoliar Borne among the leaves. **[247]**

intergeneric hybrid A plant produced by crossing species of two different genera, e.g. x *Cupressocyparis leylandii*, a hybrid between *Cupressus macrocarpa* and *Chamaecyparis nootkatensis*.

internodal Between nodes.

internode The part of the stem between two adjacent nodes. **[152, 169, 170, 184]**

interpetiolar Between the petioles.

interspecific hybrid A plant produced by crossing two species within the same genus, e.g. *Mahonia* x *media*, a hybrid between *Mahonia japonica* and *M.lomariifolia*.

intine The inner part of the wall of a pollen grain. **[212]**

intrafloral Within the flower.

intramarginal Within and close to the margin.

intrastaminal Within the stamens.

introduced Brought in from another region.

introrse With the anthers facing and opening inwards, towards the centre of the flower. **[218]**

intrusive-parietal With the placentas projecting inwards from the walls of the ovary, in some cases almost meeting at the centre.

inverted Reversed, turned upside down.

in vitro **culture** (=in glass) Studies on living material performed under sterile conditions away from the plant from which it was obtained. (includes micropropagation)

involucel The involucre of a partial umbel in the Umbelliferae; the epicalyx of connate bracteoles at the base of each individual floret in the Dipsacaceae.

involucral Relating to the involucre. **[184]**

involucrate With an involucre.

细胞环。[171]

叶间的 生于叶之间的。[247]

属间杂种 两个不同属的种杂交产生的植物，如杂扁柏（×*Cupressocyparis leylandii*），是大果柏木（*Cupress macrocarpa*）和加拿逊扁柏（*Chamaecyparis nootkatensis*）的杂种。

节间的 两个节之间。

节间 两个相邻的节之间的茎的部分。[152, 169, 170, 184]

叶柄间的 两个叶柄之间。

种间杂种 同一属内两个种杂交产生的植物，如杂种十大功劳（Mahonia×media）是十大功劳（Mahonia japonica）和阿里山十大功劳（M. lomarifolia）的杂种。

内壁 花粉粒壁的内层。[212]

花内的 在花之内。

近边缘内的 边缘内或接近边缘的。

雄蕊内的 在雄蕊之内。

引进的 从另一地区引来的。

内向的（花药） 具有面向花中心开裂的花药的。[218]

侵入侧膜胎座的 从子房壁向内突起形成的胎座，在某些情况下几乎达到中心。

倒颠的 倒转的，反向的。

体外培养（=试管培养） 在不育条件下进行的、远离该植物的活材料研究（包括微繁殖）。

小总苞 伞形科植物部分伞形花序的总苞；川续断科植物每一小花基部的合生小苞片的副萼。

总苞的 与总苞有关的。[183]

具总苞的 具有总苞的。

involucre A ring of bracts surrounding the head of flowers in the Compositae or subtending the umbel in the Umbelliferae [258]; in some species of gymnosperms, the whorl of scales subtending the cone.

involute Rolled inwards at the margin, i.e. towards the adaxial surface. [188]

irregular Not actinomorphic, sometimes applied to both zygomorphic as well as asymmetric flowers.

isobilateral Divisible into two similar halves; in leaves, having both surfaces similar to each other.

isodiametric Of cells, having equal diameters, i.e. roughly spherical in shape.

isomerous Having an equal number of members in successive series or whorls.

isomorphic Similar in form.

isophyllous Having leaves of only one kind.

isostemonous Having as many stamens as petals.

isthmus The narrow part that connects two broader parts of the same organ. [177]

jaculator A hook-like outgrowth from the stalk of the seed which aids its dispersal, as in members of subfamily Acanthoideae in the Acanthaceae. [160]

joint (see **node**).

jugate Joined together in pairs.

juvenile foliage The young leaves of e.g. *Eucalyptus* or *Juniperus* which differ in shape and colour from the mature or adult leaves. [269]

K (in a floral formula) Calyx, e.g. K5 indicates a calyx composed of 5 sepals.

keel (see **carina**).

keiki In orchids, a plantlet that develops adventitiously on a stem, pseudobulb, or branch of an inflorescence.

总苞 菊科植物头状花序周围的或伞形科植物包着伞形花序一轮苞片 [258]；某些裸子植物包着球果的那轮苞片。

内卷的 边缘内卷的，即向近轴面卷起。[188]

不整齐的 非辐射对称，有时用于两侧对称和不对称的花。

两侧相等的 分成相同两半的；两面彼此相同的叶。

等径的 细胞具等径的，即形状上粗略为球形的。

等基数的 在依次的级数或轮上具相同数目成员的。

同形态的 形状相似的。

等叶的 仅具有一种类型叶的。

同基数雄蕊的 具有与花瓣同数雄蕊的。

峡（部） 同一器官连接两个较宽部分的狭窄处。[177]

（珠）种柄钩 种柄上形成的用以传播种子的钩状生长物，见于爵床科爵床亚科植物。[160]

节，关节 （见"节"）。

成对的 成对连在一起。

幼叶 如桉属或刺柏属的与成熟叶或成年叶不同形状和颜色的幼叶。[269]

花萼（在花程式中） 花萼，例如K5表示花萼由5个萼片组成。

龙骨状突起、龙骨瓣 （见"龙骨状突起"）。

后代 在兰花中，偶然生长在茎、假球茎或花序分枝上的苗。

kernel The nucellus of an ovule or a seed; the grain of a cereal grass **[244]**; the edible part of a nut within its hard pericarp.

kex The dry, usually hollow stem of a large umbellifer, or the whole umbelliferous plant.

kingdom The taxonomic rank that includes all plants and comprises a number of divisions.

knee root One of the breathing roots that grow upwards from the submerged roots of tropical swamp plants and project above the surface of the water or mud to form an angular struchure or 'knee'. In the case of *Taxodium distichum* (Swamp Cypress) they are known as 'cypress knees'. **[147]**

knot The hard tissue formed where a branch grows out of the trunk of a tree, clearly visible in cross-section in timber.

knur, knurr A swollen outgrowth from a tree trunk.

labellum A lip, especially the highly modified third petal in the Orchidaceae. **[236, 251]**

labiate Lipped **[211]**; a member of the Labiatae.

lacerate Jagged, irregularly cut as if torn. **[193]**

lachrymiform, lacrimiform Tear-shaped, obovate.

laciniate Deeply cut into narrow lobes. **[193]**

lacuna A space or cavity. **[148]**

lacunar Relating to lacunae.

lacunate Having lacunae.

lamina (blade) The expanded part of a leaf or frond. **[179, 181, 188]**

laminar Blade-like.

lanate Woolly. **[200]**

lanceolate Lance-shaped. **[190]**

lanose, lanuginose Woolly.

lanulose Diminutive of lanose.

lateral At the side. **[148, 150, 151, 155, 226, 245, 255]**

lateral dehiscence The shedding of pollen

核，仁，谷粒 胚珠或种子的珠心；禾本科植物的谷粒 [244]；坚果坚硬内果皮内的可食部分。

干茎 大型伞形科植物的干燥、通常中空的茎或整个伞形科植物。

（植物）界 分类学等级，包含所有植物，包括很多门。

膝根 一种呼吸根，由热带沼泽植物的沉水根向上生长，在水或泥的表面伸出，形成的角状或膝状的结构。就美国水松来说，它的膝根被称为"落羽杉膝"。[147]

（树木或木材上的）节，结 树干长出树枝处形成的一种坚硬组织，在木材横切面上清晰可见。

（树）瘤，硬的突起 树干上长出的膨大生长物。

唇瓣 唇瓣，尤其是兰科植物高度修饰的第三花瓣。[236, 251]

唇形的 唇形的 [211]；唇形科一员。

撕裂状的 参差不齐的，不规则剪切像撕裂。[193]

泪滴形的 泪滴形的，倒卵形的。

条裂的 深度剪切成狭窄裂片的。[193]

腔隙 空间或腔隙。[148]

腔隙的 与空间或腔室相关的。

间隙的 具有空间或腔室的。

叶片 一枚叶的扩展部分。[179, 181, 188]

片状的 叶片状的。

具绵状毛的 羊毛状的。[200]

披针形的 矛形的。[190]

具绵状毛的 羊毛状的。

短绵毛状的 小绵毛状的。

侧面的 在侧面的。[148, 150, 151, 155, 226, 245, 255]

侧面开裂 花粉粒从花药侧面的裂缝中

from a split at the side of the anther.

lateral growth Increase in girth of a stem or root, resulting from the activity of the cambium.

late wood (see **antumn wood**).

latex A juice produced by special cells in many different plants. It is usually milky (as in *Taraxacum*, dandelion) but may be colourless, yellow, orange (as in *Chelidonium majus*, Greater Celandine) or red. In some tropical trees, e.g. *Hevea brasiliensis*, the latex can be collected and processed to form rubber.

laticiferous Bearing latex.

latiseptate Having the partition (septum) across the broadest diameter of the fruit, as in *Lunaria annua* (Honesty). **[265]**

latrorse With the anthers facing and opening sidewards, as in *Begonia cucullata*.

lax Loose or open, not dense.

leaf A lateral outgrowth from the stem, usually consisting of a stalk (petiole) and a flattened blade (lamina). **[186—197]**

leaf blade (see **lamina**).

leaflet A leaf-like segment of a compound leaf. **[180, 192]**

leaf mosaic The overall arrangement of the leaves of a plant, determined by their shape and by phyllotaxy, which allows the maximum amount of light to flall upon each leaf. **[175]**

leaf-opposed Borne on the stem but on the opposite side from a leaf, as the tendrils in some species of vines, e.g. *Parthenocissus*. **[179]**

leaf rosette A circular cluster of leaves at the base of a stem. **[175]**

leaf sheath The lower part of a leaf stalk which more or less encloses the stem. **[187]**

leaf trace The vascular tissue that leads from the vascular system of the stem to the base

散出。

侧面生长 形成层活动引起的茎或根周长的增加。

晚材 （见"秋材"）。

乳汁 许多不同植物的特殊细胞产生的汁液。它通常是乳汁状的（像蒲公英），但可以是无色的、黄色的、橘红色的（像白屈菜）或红色的。一些热带树种，如巴西橡胶树，它能被采集、加工形成橡胶。

具乳汁的 产生乳汁的。

具宽隔膜的 有经过果实宽径的隔膜，像银扇草。[265]

侧向纵裂的 具侧裂花药的，像四季秋海棠。

疏松的 疏松的或散开的，不密集的。

叶 一类茎的侧面生长物，通常包括叶柄和扁平的叶片。[186—197]

叶片 （见"叶片"）。

小叶 复叶的一个叶状部分。[180, 192]

叶镶嵌 由植物叶的形状和叶序决定的、允许最大量光线落到每一片叶上的叶的总体排列方式。[175]

与叶对生的 生在茎的与叶相对的一侧上，像藤本植物某些种的卷须，如爬山虎属。[179]

叶莲座状 茎基部叶呈莲座状的。[175]

叶鞘 或多或少抱茎的叶柄的下部。[187]

叶迹 引导茎内维管系统到叶基部的维管组织。[168]

of a leaf. **[168]**

leg. (legit) Collected by, used on a herbarium label and followed by the name of the collector of the specimen concerned.

legume The two-valved fruit formed from a single carpel in most members of the Leguminosae. **[219]**

leguminous Bearing legumes.

lemma The lower of the two bracts enclosing a grass flower. **[243]**

lenticel A pore in the stem that allows gases to pass between the outside atmosphere and the interior of a plant. **[152, 170]**

lenticular (biconvex) Lens-shaped, convex on both sides. **[160]**

lepidote Covered with small, fine scales. **[199]**

leptocaul Having a relatively slender woody stem.

leucoplast A colourless plastid that in roots, tubers, and other underground parts of plants can convert sugar into starch and is then termed an amyloplast. It is often capable of developing chlorophyll, as when potato tubers turn green in the presence of light.

liana, liane A woody climber in tropical forests that grows from the ground into the tree canopy.

lid The more or less flat structure that is attached to one side of the mouth of the pitcher-shaped leaves that act as insect traps in species of *Nepenthes*. **[181]**

life cycle The course of development from any given stage in the life of a plant until the same stage is reached again.

ligneous Woody.

lignified Converted into wood. **[171]**

lignin A hard substance found in the thickened cell walls of xylem and sclerenchyma.

ligulate Strap-shaped. **[190, 211]**

ligule A strap-shaped structure, such as the limb of the ray florets in the Compositae

采集人 用在标本标签上，后面紧接相关标本的采集者姓名。

荚果 豆科绝大多数植物的由单心皮子房形成的两果瓣的果实。 [219]

荚果的 生有荚果的。

外稃 包住禾草类植物花的两苞片中下面的一枚。 [243]

皮孔 允许气体在外部空气和植物体内部之间进出的茎上小孔。 [152, 170]

透镜状的 透镜状的，两面凸起的。 [160]

具鳞片的 覆有细小鳞片的。 [199]

具细茎的 具相对纤细木质茎的。

白色体 在植物根、块茎和其他地下器官中，将糖类转化为淀粉的无色质体，故被称为造粉体。它通常能发育为叶绿体，如马铃薯块茎在阳光存在时变绿。

藤本植物 热带雨林中，从地面生长到树顶的木质攀缘植物。

盖 附着于囊状叶口部一侧、起诱捕昆虫作用的、或多或少扁平的结构，如猪笼草等某些植物。 [181]

生活史，生活周期 在植物生活中，从任一给定阶段开始，到这一阶段再次到达时结束的发育过程。

木质的 木质。

木质化的 转变成木质的。 [171]

木质素 一类在木质部和机械组织的加厚细胞壁中发现的坚硬物质。

舌状的 舌状的。 [190, 211]

舌片 舌状结构，如菊科植物舌状花的舌片 [234]；禾草类植物叶鞘顶端形成的干膜

[234]; the scarious projection from the top of the leaf sheath in grasses. **[187]**

liliaceous Resembling a lily flower. **[211]**

Liliopsida (see **monocotyledons**).

limb The broadened upper part of a separate petal, as distinct from the claw **[227]**; the spreading rim of a gamopetalous flower, as distinct from the tube, as in *Primula veris*. **[229]**

limbate With a distinct edge or rim, especially when of a different colour from the inner part.

limen The rim at the base of the androgynophore in the genus *Passiflora* (passion flower). **[228]**

linear Long and narrow with parallel sides. **[190, 221]**

lineate Marked with lines.

lingulate Tongue-shaped. **[190]**

Linnaean Relating to the Swedish biologist Linnaeus (Carl von Linné, 1707-78), who developed a system of classification for plants and animals involving binomial nomenclature.

lip One of the two divisions of a bilabiate corolla or calyx **[208]**; the labellum in the Orchidaceae. **[251—253]**

lithocyst A cell containing a cystolith.

lithophile (see **lithophyte**).

lithophilous (see **lithophytic**).

lithophyte (lithophile) A plant growing amongst or on rocks or on cliff faces.

lithophytic (lithophilous) Growing amongst or on rocks or on cliff faces.

littoral Relating to the seashore or lakeside.

lobate (lobed) Having one or more lobes. **[221]**

lobe Any division of an organ, especially if the part is rounded. **[194]**

lobed (see **lobate**).

lobulate Having small lobes.

质的突起物。 [187]

象百合的 像百合花的。 [211]

百合纲 （见单子叶植物）。

瓣片，冠檐 一枚离生花瓣的上面加宽部分，与爪不同 [227]；合瓣花的辐射开展的檐部，与花冠筒不同，如黄花九轮草。 [229]

具异色边的 具明显边缘的，尤其是与内面颜色不同时。

蕊柄盘 西番莲属植物雌雄柄基部的边缘。 [228]

线形的，条形的 细长狭窄具平行边的。 [190, 221]

具线纹的 以线条标记的。

舌状的 舌形。 [190]

林奈的或林奈分类学和命名法的 与瑞典生物学家林奈有关的，他提出了关于双名命名法的植物和动物分类系统。

唇瓣 二唇形花冠或花萼的上下两个裂片之一 [208]；兰科植物的唇瓣。 [251—253]

晶细胞 一类含有钟乳体的细胞。

石生植物 （见"石生植物"）。

石生植物的 （见"石生植物的"）。

石生植物 生长于岩石或峭壁间或岩石或峭壁上的植物。

石生植物的 生长于岩石或峭壁间或岩石或峭壁上的。

海岸的，湖边的 与海边、湖边相关的。

具裂片的 具一或多枚裂片的。 [221]

圆裂片 一个器官的任一部分，尤其是若这个部分是圆形的。 [194]

分裂的 （见"分裂的"）。

具小裂片的 具有小的裂片。

lobule A small lobes.

loc. class. (locus classicus) Classical locality, i.e. the place from where the plant concerned was originally collected.

locule (see **loculus**).

loculicidal Splitting at maturity into the loculus, more or less midway between the partitions of the capsule. **[219]**

loculus (locule) A compartment of an ovary or an anther. **[213, 219]**

lodicule One of usually two minute scales in a grass flower, generally considered to be a vestigial perianth. **[243]**

loment (see **lomentum**).

lomentose Relating to a lomentum.

lomentum A fruit derived from a single carpel which breaks up into one-seeded portions, as in the genus *Ornithopus* (Leguminosae). [255]

long-day plant A plant which needs prolonged periods of light alternating with shorter periods of darkness for the proper development of its flowers and fruit.

longitudinal Lengthwise.

longitudinal dehiscence The shedding of pollen from a lengthwise split in the anther, or seeds from a lengthwise split in a capsule. **[256]**

lorate Strap-shaped, usually more broadly than in ligulate. **[190]**

L.S. Longitudinal section.

lunate Crescent-shaped. **[191]**

lupulin A secretion from the glands present on the fruits (strobiles) of *Humulus lupulus* (Hop). **[262]**

lustrous (glossy) Smooth and shiny.

lyrate Pinnatifid, with the terminal lobe rounded and much larger than the others. **[191]**

m Metre(s).

小裂片 小裂片。

采集地 原采集地点，即被关注的植物首次采集到的地区。

室 （见"室"）。

室背开裂的 蒴果成熟时或多或少从两隔膜间中部裂成数室。 [219]

室 子房或花药的内部空间。 [213, 219]

浆片 禾草类植物花的两个小鳞片之一，一般被认为是残留的花被。 [243]

节荚 （见"节荚"）。

节荚的 与节荚有关的。

节荚 由单一心皮子房发育而来的、成熟时断裂为含一枚种子的部分，如鸟足豆属（豆科）。 [255]

长日照植物 一类需要延长光照时间和较短黑暗时间相交替才能使其花果正常发育的植物。

纵向的 纵长的。

纵裂 花粉粒从花药的纵向裂缝中散出；种子从蒴果的纵向裂缝中散出。 [256]

带状的 带状的，通常比舌状的更宽。 [190]

纵剖面。

新月形的， 新月形的。 [191]

啤酒花苦味素 啤酒花果实（果穗）上的腺体产生的分泌物。 [262]

有光泽的 光滑而有光泽的。

丈头羽裂的 羽状裂的，具圆形、较其他裂片大得多的顶生裂片。 [191]

米 米。

macrogamete (see **megagamete**).

macronutrients The elements nitrogen, phosphorus, potassium, calcium, magnesium, and sulphur, that are required in relatively large amounts for the formation of plant tissue. (see also **micronutrients**)

macrophyll The term proposed by Raunkiaer for leaves of a large size.

macrophyllous (see **megaphyllous**).

macrophyte A plant, especially an aquatic plant, large enough to be visible to the naked eye.

macrospore (see **megaspore**).

macrothermic (see **megathermic**).

maculate Having spots or markings.

Magnoliophyta (see **Angiospermae**).

Magnoliopsida (see **dicotyledons**).

male flower Having fertile stamens, but only a rudimentary or non-functional gynoecium. **[207, 208, 227, 263]**

male nucleus (sperm cell) One of the two cells into which the generative cell divides, at the time of or sometimes before pollination. **[213]**

malodorous Having an unpleasant smell.

many-n Polyploid.

marcescent Withering, but not falling off. **[247]**

marginal placentation The arrangement in which the placenta extends along one side of the ovary of a free carpel. **[222, 223]**

massula A cluster of pollen grains developed from a single cell. **[281]**

mast The fallen fruit of certain genera in the Fagaceae, e.g. *Fagus* (beech) and *Quercus* (oak), formerly used as food for animals.

matted Tangled into a dense mass.

mealy Covered with a coarse, flour-like powder.

medial, median Relating to the middle. **[251]**

medifixed Attached at the middle.

大配子 （见"大配子"）。

大量元素 在植物组织形成中需要的相对大量的元素N、P、K、Ca、Mg和S。（见"微量元素"）。

大型叶 瑙基耶尔为大型叶而提出的术语。

大型叶的 （见"大型叶的"）。

大型植物 大到肉眼可见的植物，特别是水生植物。

大孢子 （见"大孢子"）。

高温植物的 （见"高温植物的"）。

具斑点的 具点的，或斑点的。

木兰植物门 （见"被子植物"）。

木兰植物纲 （见"双子叶植物"）。

雄花 具可育雄蕊，但仅有未完全发育或无功能雌蕊群的花。 [207, 208, 227, 263]

雄核（精子细胞） 生殖细胞在传粉之前分裂形成的两个细胞之一。 [213]

恶臭的 有一种难闻的气味。

多倍体的 多倍体的。

凋存的 蔫萎，但不落下。 [247]

边缘胎座 胎座沿一枚离生心皮的子房一侧边缘扩展开来的排列方式。 [222, 223]

花粉小块 从单个细胞发育而来的一簇花粉粒。 [281]

坚果，壳斗 壳斗科某些属，如青冈栎属植物（山毛榉）和栎属（橡树）植物落下的果实，曾用作动物饲料。

蓬松的，散乱的 缠结成密集的团块。

粉状的 覆盖有粗糙的、面粉状的粉状物。

中间的 与中间有关的。 [251]

中着的 在中部着生的。

mediseptate Having the partition (septum) across the middle of a more or less terete fruit, as in *Aubrieta deltoidea*. **[265]**

medulla The central part of an organ, the pith in a young stem. [171]

medullary ray (pith ray) One of the sheets of tissue in the stem of a dicotyledon that extend from the medulla or pith to the cortex (primary medullary ray), or outwards and inwards from the vascular cambium (secondary medullary ray). **[171]**

megagamete (macrogamete) In an organism where male and female gametes differ in size, the larger, usually female gamete.

megaphyll The term proposed by Raunkiaer for leaves of the largest size.

megaphyllous (macrophyllous) Large-leaved.

megasporangiate Relating to a megasporangium.

megasporangium The structure in which megaspores are produced. **[284]**

megaspore A spore that develops into a female gametophyte. **[284]**

megasporophyll A specialised leaf in heterosporous plants that bears the megasporangia; in gymnosperms, one of the ovuliferous scales that are arranged round the central axis of a female cone, and correspond to the carpels in a flowering plant. **[267, 272]**

megathermic Requiring much heat for growth and development, as tropical plants.

membranous Thin and semi-transparent like a membrane. **[151]**

mentum In some orchids, e.g. *Dendrobium*, a chin-like projection formed by the base of the column and the lateral sepals. **[253]**

mericarp A portion of a schizocarp which splits away at maturity as a perfect fruit, as in most members of the Geraniaceae and Umbelliferae. **[232, 257]**

meristele One of the strands of vascular

中纵隔膜 或多或少圆柱状果实中部具纵向隔膜的，像紫芥菜的果实。[265]

中部，髓部 器官的中部；幼茎的髓部。[171]

髓射线 双子叶植物茎内从中部或髓部到皮层展开的薄壁组织（初级髓射线），或从维管形成层向外向内形成的射线（次级髓射线）之一。[171]

大配子 生物体的雄配子和雌配子在体积上大小不同，大的、通常为雌性的配子。

大型叶 瑙基耶尔为最大型的叶提出的术语。

大型叶的 大叶的。

大孢子囊的 与大孢子囊有关的。

大孢子囊 产生大孢子的结构。[284]

大孢子 将来发育成雌配子体的孢子。[284]

大孢子叶 异型孢子植物的能够产生大孢子囊的一种特化叶；裸子植物的围绕雌球果中轴排列的珠鳞之一；有花植物中相应的心皮。[267, 272]

高温植物的 像热带植物一样，为生长和发育需要很多热量的。

膜质的 薄而半透明，像膜的。[151]

萼囊 某些兰花，如石斛属植物花中，从合蕊柱和侧生萼片基部形成的囊状突起物。[253]

分果爿，分果瓣 成熟时作为完整果实裂开的分果部分，如在牻牛儿苗科和伞形科的绝大多数植物。[232, 257]

分体中柱 在内皮层以内，组成网状中柱

tissue, consisting of xylem surrounded by phloem within a sheath of endodermis, that compose a dictyostele. **[280]**

meristem An area of tissue, found especially in the tips of shoots and roots, and in the cambium, that continues to undergo cell-division throughout the life of the plant. **[173]**

meristematic Relating to a meristem.

mesocarp The middle layer of the pericarp. **[259]**

mesochile The middle portion of the labellum in an orchid. **[253]**

mesophyll The parenchymatous, photosynthetic tissue that forms the inner part of the leaf blade between the upper and lower epidermis. In many dicotyledons the upper part consists of the palisade mesophyll and the lower part the spongy mesophyll. **[196, 268]**

mesophyte A plant that is adapted to grow in a moist habitat, where there is no prolonged drought.

mesophytic Growing in moist habitats.

mesotherm A plant requiring moderate heat for its optimal growth and development.

mesotonic A type of branching in which the shoots nearest the middle of the stem show the greatest development.

metagyny (see **protandry**).

metamorphosis Transformation of one structure into another, as stamens into petals; a change in the type of branching of a tree from generally plagiotropic to generally orthotropic.

metaphloem The phloem that is formed after the protophloem.

metaxylem The xylem that is formed after the protoxylem.

microclimate The climatic conditions existing in a small, localised area.

microgamete In an organism where male and female gametes differ in size, the smaller,

的、由木质部及周围环绕的韧皮部构成的维管组织束之一。[280]

分生组织 特别是在茎尖和根尖及形成层中发现的、植物生命活动中持续进行细胞分裂的一个组织区。[173]

分生组织的 与分生组织有关的。

中果皮 果皮的中间层。[259]

唇瓣中部 兰花唇瓣的中间部分。[253]

叶肉 一类形成叶片上、下表皮之间内部部分的薄壁的光合组织。在许多双子叶植物中，其上部包括栅栏组织而下部包括海绵组织。[196, 268]

中生植物 一类适于生长在没有持续干旱的潮湿生境的植物。

中生植物的 生长于潮湿生境的。

中温植物 为保证植物最佳生长和发育，需要中等热量的植物。

中部优势 茎最接近中部的分枝显示出最大发育优势的一种分枝类型。

雌蕊后熟（见"雄蕊先熟"）。

变态 一种结构转变为另一种结构，如雄蕊转变为花瓣；一般从斜生到直生的树木分枝类型的一种改变。

后生韧皮部 原生韧皮部之后形成的韧皮部。

后生木质部 原生木质部之后形成的木质部。

小气候 存在于小的、局限性区域的气候条件。

小配子 生物体的雄配子和雌配子在体积上大小不同，小的、通常为雄性的配子。

usually male gamete.

micronutrients (trace elements) The elements that, together with the macronutrients, are necessary for the successful growth of a plant, but which are required only in small amounts. They include iron, manganese, boron, zinc, molybdenum, chlorine, and copper.

micropropagation The development of new plants from very small pieces of plant tissue, e.g. embryos, shoot tips, root tips etc., in an artificial medium and under sterile conditions (included in the concept of *in vitro* culture).

micropylar Relating to the micropyle. **[159]**

micropyle The opening in the integuments of an ovule, through which the pollen tube grows after pollination. **[157, 213, 223]**

microspecies Species founded on minute differences and used mostly for apomictic plants, as in *Taraxacum* (dandelion) and *Hieracium* (hawkweed).

microsporangiate Relating to a microsporangium. **[273]**

microsporangium The spore sac in which microspores are produced. **[267, 271, 272]**

microspore A spore that develops into a male gametophyte, corresponding to a pollen grain in a flowering plant. **[284]**

microsporophyll A specialised leaf in heterosporous plants that bears the microsporangia; in gymnosperms, one of the bracts that are arranged round the axis of a male cone, and correspond to the stamens in a flowering plant. **[267, 272]**

midrib The middle and principal vein of a leaf. **[188, 272]**

mm Millemetre(s).

monad A single pollen grain, not united with others.

monadelphous Having the stamens united in

微量元素（痕量元素） 和大量元素一起对植物成功生长都必不可少，但仅需要少量的元素。这些元素包括铁、锰、硼、锌、钼、氯和铜。

微繁殖 在人工培养基和无性条件下，从很小的一块植物组织如胚、枝尖和根尖等发育成一株新植物的繁殖方式（包括在"在实验室条件下培植"的概念中）。

珠孔的 与珠孔相关的。 [159]

珠孔 传粉后花粉管生长、进入胚珠经过的珠被的开口。 [157, 213, 223]

小种 大多用于单性生殖植物、建立在微区别之上的种，如蒲公英属（蒲公英）和山柳菊属（山柳菊）。

小孢子囊的 与小孢子囊有关的。 [273]

小孢子囊 产生小孢子的孢子囊。 [267, 271, 272]

小孢子 将来发育成雄配子体，有花植物相应的花粉粒的孢子。 [284]

小孢子叶；雄蕊 异型孢子植物的能够产生小孢子囊的一种特化叶；裸子植物的围绕雄球果中轴排列的珠鳞之一；有花植物中相应的雄蕊。 [267, 272]

中脉，中肋 叶的中脉或主脉。 [188, 272]

毫米 毫米。

单花粉 单粒花粉，不与其他结合的。

单体雄蕊的 通过花丝融合而结合成为一组

one group by the fusion of their filaments, as in some of the Leguminosae, Polygalaceae and Malvaceae. **[215]**

monandrous Having one stamen, as most orchids.

monarch root A root with a single protoxylem strand in the stele.

moniliform Like a string of beads. **[146, 202]**

monocarpellary (see **monocarpous**)

monocarpic (hapaxanthic) Flowering and fruiting once only before dying.

monocarpous (monocarpellary) Of a fruit, composed of a single carpel.

monocaulous Having one stem.

monocephalic Bearing one head of flowers, as the scape of plants in the genus *Taraxacum* (dandelion).

monochasial cyme, monochasium A cyme with lateral branching on one side only of the main axis. **[204]**

monochlamydeous (haplochlamydeous) Having a perianth of a single whorl, i.e. either the calyx or the corolla is present.

monoclinous (see **hisexual**)

monocolpate Of a pollen grain, having a single colpus.

monocotyledons (Monocotyledonae, Liliopsida) Flowering plants having one cotyledon. **[168]**

monocotyledonous Having one cotyledon.

monoecious With male and female flowers on the same plant.

monoecism, monoecy The state of being monoecious.

monogeneric Of a family, containing only a single genus.

monograph A systematic account of a particular genus or family.

monomerous Formed of a single unit, as a monocarpous fruit.

monomorphic Occurring in only one form.

的雄蕊，如豆科、远志科和锦葵科的某些植物。[215]

单雄蕊的 具有一枚雄蕊，如绝大多数的兰花。

单原型根 一类中柱中仅具单个原生木质部束的根。

念珠状的 像一串念珠。[146, 202]

单心皮的（见"单心皮果的"）。

结一次果的（开一次花的） 仅在死亡前一次开花结果的。

单心皮果的 由一心皮构成的果实的。

单茎的 具有一条茎的。

单头的 生有一个头状花序的，像蒲公英属植物（蒲公英）的花葶上。

单歧聚伞花序 在花序主轴上仅具一侧侧枝的聚伞花序。[204]

单倍花的 仅具一轮花被的，即花萼存在或花冠存在。

雌雄同花的（见"两性的"）。

单沟的 具单一萌发沟花粉粒的。

单子叶植物（单子叶植物纲，百合纲） 具一枚子叶的有花植物。[168]

单子叶的 具一枚子叶的。

雌雄同株的 雌花和雄花在同一植株上的。

雌雄同株 雌雄同株的状态。

单属的 仅含单一属的科。

专著 一种特定属或科的系统的记述。

单基数的 由单一单位形成的，像一心皮果实。

单型的 仅以一种形式发生的。

monopetalous With only one petal; gamopetalous.

单瓣的 仅具一枚花瓣的；合瓣的。

monopodial With a simple main stem or axis, growing by apical extension and bearing lateral branches.

单轴分枝的 具单一主茎或主轴的，通过顶端延伸生长，具有侧枝。

monoporate Of a pollen grain, having one pore.

（花粉粒）单孔的 花粉粒具一个萌发孔的。

monospecific Of a genus, containing only a single species.

单种属 仅具单种的属。

monosulcate Of a pollen grain, having one groove or furrow.

（花粉粒）单槽的 花粉粒具一条萌发沟的。

monotelic With each lateral inflorescence-branch ending in a flower, as a cyme. **[204]**

单顶的 每一侧生花序分枝顶端都止于一朵花，如聚伞花序。[204]

monothecous With one anther cell.

单囊的 具有一个花粉囊的。

monotypic Having only one representative, as a genus with one species.

单型的，单种的 仅有一个代表的，像一个属只有一个种。

montane Growing in mountainous regions.

山地的，山上生的 生长于山区的。

morphological Relating to the form of a plant.

形态学的 与植物形态有关的。

morphology The science or study of the form of plants, as distinct from anatomy.

形态学 一门与解剖学明显不同的、研究植物形态的科学。

mother bulb A mature bulb that is capable of flowering and producing one or more bulblets (daughter bulbs). **[168]**

母鳞茎 一类能开花，并能产生一或多个子鳞茎的成熟鳞茎。[168]

mother cell A cell that divides into two daughter cells.

母细胞 能够分裂成两个子细胞的细胞。

motile Capable of independent movement.

能动的 能独立运动的。

mucilage A sticky substance or solution.

黏液，胶浆 一类黏的物质或液体。

mucilaginous Slimy.

黏质的 黏的。

mucro A short straight point. **[195]**

短直尖。[195]

mucronate Ending abruptly in a short, straight point. **[195]**

（叶子等）具尖的 （叶子等）末端突然变成短而直的尖的。[195]

mucronulate Ending abruptly in a very short, straight point. **[195]**

具小短尖的 末端突然变成非常短而直的尖的。[195]

multiaperturate Of a pollen grain, having many pores.

多个萌发的 花粉粒具多萌发孔。

multicellular Many-celled. **[202]**

多细胞的 具多细胞的。[202]

multiciliate Having many marginal hairs.

多纤毛的 具有许多边缘毛的。

multilocular With many loculi or cells.

多室的 具有许多室的。

multiple fruit (collective fruit) A fruit formed from an inflorescence, often including

复果，聚花果 由一枚花序形成的果实，通常含苞片，像菠萝属植物（菠萝）和桑

bracts, as in *Ananas* (pineapple) and *Morus* (mulberry). **[262]**

multiseriate In several series, rows or whorls.

muricate Rough with short, hard points. **[199]**

mutant (sport) An individual that has arisen as a result of mutation.

mutation A genetic change that may occur spontaneously or may be induced artificially by the use of certain chemicals; a mutant.

mutualism A form of symbiosis in which two different organisms co-exist to their mutual advantage.

mycorrhiza The association of fungi and the roots of plants to their mutual advantage. **[282]**

mycorrhizal Relating to mycorrhiza.

mycotroph A plant that lives in symbiosis with a fungus.

myrmecochorous Having seeds that are dispersed by ants.

myrmecochory Dispersal of seeds by ants.

myrmecophilous Having the stem or root inhabited by ants. **[182]**

myrmecophily Symbiosis between ants and plants; pollination by ants.

n Haploid; a figure preceding the 'n' indicates the number of sets of chromosomes in each cell, e.g. 2n=diploid.

n- (notho) A prefix placed before a taxonomic rank to indicate a hybrid.

N (see **nitrogen**).

NaCl (see **sodium chloride**).

naked Lacking a covering, as a flower without a perianth, e.g. *Salix* (willow). **[211]**

napiform Turnip-shaped, e.g. the hypocotyl of *Brassica rapa* (Turnip). **[153]**

narrowly upright With the branches growing up more closely to the trunk. **[163]**

nastic movement A plant movement

属植物（桑）。 [262]

多列的 多列的、多行的或多轮的。

粗糙的 粗糙的，具短小、坚硬尖头的。 [199]

突变体 一类作为突变结果而发生的个体。

突变 一类自然发生的或通过应用某些化学物质人工诱导的基因变化；突变体。

互利共生、互利共栖 两个不同的生物互利共存的一种共生关系形式。

菌根 真菌和植物根的互惠结合。 [282]

菌根的 与菌根有关的。

菌根营养的植物 一类和真菌类共生的植物。

蚁传布的 具靠蚂蚁传播的种子的。

蚁布 靠蚂蚁传播种子。

适蚁的，蚁喜 具有居住着蚂蚁的茎和根的。 [182]

蚁媒 蚂蚁与植物之间的共生现象；蚂蚁传粉。

单倍体 单倍体；"n"前面的数字表示每个细胞内染色体的套数，如2n=二倍体。

杂种 放在一个分类等级前的前缀，表示杂种。

氮 （见"氮"）。

氯化钠 （见氯化钠）。

裸露的 缺乏覆盖物的，像一朵花没有花被，如柳属植物。 [211]

（根部）芜菁状的 芜菁状的，如芜菁的下胚轴。 [153]

（分枝）上拢的 具向上更靠近树干生长的树枝的。 [163]

感性运动 一种不依靠直接的外部刺激的植

independent of the direction of the external stimulus, such as the opening or closing of some flowers in response to an alteration in temperature or light intensity. [174]

natant Floating in water.

native (see **indigenous**).

naturalised Thoroughly established after introduction from another region.

navicular (cymbiform) Boat-shaped, as the united lower petals in flowers of subfamily Papilionoideae in the Leguminosae. [230]

nec Nor, nor of.

necrosis Localised death of cells which leaves the surrounding plant tissue unaffected.

nectar A sugar solution, attracting insects or birds to flowers for the purpose of pollination. [207]

nectar guide (honey guide) Marking on the perianth of a flower, usually consisting of lines or dots, that direct a pollinating insect to the nectary. [231]

nectariferous Bearing nectar-secreting glands. [177]

nectar pit A depression in which nectar collects.

nectar roll The nectary in carnivorous plants such as *Sarracenia*, in which the edge of the leaf forming the mouth of the pitcher-like trap is rolled over. [181]

nectary (honey gland) A gland or surface from which nectar is secreted. [226, 231]

needle A long, narrow leaf, characteristic of many conifers. [267—269]

nervate Nerved or veined. [189]

nervation, nervature (see **venation**).

nerve (see **vein**).

nervure A principal vein of a leaf. [189]

net-veined Having veins that join together to form a network across the lamina. [189]

neuter Having both stamens and gynoecium

物运动，如某些花的开放是对温度和光照强度改变的响应。 [174]

浮游的 在水面上漂浮的。

（植物等）本地生的，土著的 （见"乡土的、土著的"）。

归化的 从另一地区引种后已全部定居稳定生长的。

舟状的 船状的，像豆科蝶形花亚科植物花的联合下花瓣。 [230]

也不，果然不，的确不 也不，也不是的。

坏死，枯斑 细胞的局部性死亡，其周围的植物组织不受影响。

花蜜 一类引诱昆虫或鸟类到达花上进行传粉的糖液。 [207]

蜜腺标记 花被上指导传粉昆虫到达蜜腺的标记，通常包括线条或斑点。 [231]

分泌蜜汁的 生有分泌花蜜的腺体。 [177]

花蜜坑 采集蜜汁的凹陷处。

花蜜卷（辊） 肉食植物如瓶子草形成捕虫囊口部的叶边缘外卷的蜜腺。 [181]

蜜腺 分泌蜜汁的腺体或表面。 [226, 231]

针叶 长、狭窄的叶许多球果类植物的特征。 [267—269]

具脉的 有叶脉的或有脉纹的。 [189]

脉序 （见"脉序"）。

脉 （见"脉"）。

主脉，（叶）脉 叶的主脉。 [189]

具网状脉的 具有连接起来在叶片上形成网格的叶脉的。 [189]

中性的 具无功能雄蕊和雌蕊的。

non-functional.

neutral soils Soils with a pH value of about 7.

nitrogen (N) The most abundant gas in the atmosphere and one of the six macronutrients that, together with carbon, oxygen, and hydrogen, comprise the group of elements essential to the life of a plant.

nitrogen fixation A prosess carried out by bacteria which convert gaseous nitrogen into compounds that can be assimilated by the plant.

nocturnal Occurring at night.

nodal Relating to a node.

nodding (see **nutant**).

node (joint) The point on a stem where one or more leaves are borne. **[152, 169]**

nodose Knobby.

nodulation The formation of nodules on the roots of certain plants. **[146]**

nodule A small, rounded structure on the roots of plants, especially those of the Leguminosae that contain nitrogen-fixing bacteria. **[146]**

nodulose Bearing nodules. **[146]**

nom. cons. (nomen conservandum) Conserved name, a plant name invalid under the rules of the I.C.B.N. but which has been retained in order to avoid further (possibly wide-ranging) changes in the nomenclature.

nomenclature The naming of plants, especially the precise usage formulated in the I.C.B.N.

nom. illeg. (nomen illegitimum) Illegitimate name, a plant name invalid under the rules of the I.C.B.N.

nom. nud. (nomen nudum) Bare name, a new name published without any description of the plant in question, and therefore invalid under the rules of the I.C.B.N.

non Not, not of.

nonaploid (9n) Having nine sets of

中性土壤 pH值大约为7的土壤。

氮 空气中含量最丰富的气体，六种大量元素之一，和碳、氧、氢一起组成对植物生活起重要作用的元素。

固氮 细菌进行的、将气态氮转变成能被植物同化的化合物的过程。

夜开性的 夜间发生的。

节的 与节有关的。

下垂的 （见"下垂的"）。

节 茎上一或多枚叶着生的点。[152, 169]

具节的 具多节的。

结瘤 某些植物根上根瘤的形成过程。[146]

根瘤 植物根上的小圆形结构，特别是那些豆科植物根上的含固氮菌的结构。[146]

具根瘤的 生根瘤的。[146]

保留名 国际植物命名法规条款下无效的、为避免命名上进一步（可能更宽范围的）变化而予以保留的植物名称。

命名法；名称 植物的命名，尤其注意国际植物命名法规规定的准确用法。

不合法名称 国际植物命名法规条款下无效的植物名称。

裸名 新名称发表时在问题中没有任何植物描述，因此在国际植物命名法规条款下是无效的。

不，非，无 非，不的。

九倍体（9n） 每一细胞有九套染色体。

chromosomes in each cell.

non-endospermic seed A seed in which the nutritive tissue is absorbed more rapidly into the developing embryo so that the process is completed by the time of germination, e.g. *Vicia faba* (Broad Bean) and *Phaseolus vulgaris* (French Bean). **[157]**

non-vascular Not containing vessels.

nose The pointed end of a bulb. **[151]**

nothogenus The rank given to a hybrid genus produced by the crossing of two or more different genera, and usually indicated by a multiplication sign placed before the name, e.g. x *Cupressocyparis* (*Cupressus* x *Chamaecyparis*).

nothotaxon A unit of classification for hybrid plants. The highest rank permitted by the I.C.B.N. is nothogenus. Names of all hybrid ranks consist of the normal taxonomic rank prefixed by 'notho'.

nothospecies The rank given to a hybrid species produced by the crossing of two or more species within the same genus, and usually indicated by a multiplication sign placed before the specific epithet, e.g. *Viburnum* x *bodnantense* (*V. farreri* x *V. grandiflorum*).

nucellus The mass of cells within an ovule that contains the embryo sac. After fertilisation, it may either be absorbed or may persist to form a perisperm. **[223]**

nuciferous Bearing nuts.

nuciform Nut-like in form.

nucleus The usually ovoid or spherical structure in a plant cell that contains the chromosomes. **[212, 213]**

nucule A nutlet. **[258]**

numerous Many, often indefinite in number.

nuptial nectary A neetary situated near to the reproductive organs of the flower.

nut A dry, one-seeded, indehiscent fruit with a

无胚乳种子 一类营养组织更快地被吸收到发育的胚，以致萌发时即已完成的种子，如蚕豆和菜豆。[157]

无维管束的 不含导管。

（鳞茎）茎端 鳞茎的尖端。[151]

杂交属 两个或多个不同属间杂交产生的给予杂交属的等级，通常用名称前放一乘号"×"表示，如杂扁柏属（×*Cupressocyparis*）（柏木属*Cupressus* × 扁柏属*Chamaecyparis*）。

杂种类群 杂交植物的分类单位。国际植物命名法规许可的最高等级是杂交属。所有杂交等级的名称是正常分类等级加前缀"notho"。

杂种 同一属内两个或多个种杂交产生的给予杂种的等级，通常用种加词前放一乘号"×"表示，如博得南特荚蒾*Viburnum* × *bodnantense*（香荚蒾*V. farreri* × 大花荚蒾*V. grandiflorum*）。

珠心 胚珠内含有胚囊的细胞群，受精后它可以被吸收，也可以继续形成外胚乳。[223]

具坚果的 生有坚果的。

坚果状的 形状上坚果状的。

细胞核 植物细胞内含有染色体、通常卵形或球形的结构。[212, 213]

小坚果 小坚果。[258]

大量的 许多，通常数量不确定。

婚蜜腺 位于花生殖器官附近的蜜腺。

坚果 一类干燥、不开裂、具有木质果皮的

woody pericarp. **[184, 258]**

nutant (drooping, nodding) Bending over and pointing downwards. **[163, 164]**

nutlet A small nut, sometimes applied to an achene or part of a schizocarp. **[258]**

nutrient One of the subsrances necessary for plant growth and development. (see **macronutrients** and **micronutrients**)

nyctinasty (sleep movement) The response of plant parts, especially flowers and leaves, to night-time darkness. **[174]**

obconical (see **turbinate**).

obcordate Inversely heart-shaped, broadest towards the emarginate apex and tapering to the stalk. **[191]**

obdeltoid Triangular, with the apex truncate and tapering to the stalk. **[190]**

obdiplostemonous Having the stamens in two whorls, the inner whorl opposite the sepals and the outer whorl opposite the petals. **[214]**

oblanceolate Inversely lanceolate, broadest towards the apex and tapering to the stalk. **[190]**

oblate (see **subglobose**).

oblique Unequal, as a leaf with one side extending below the other; ascending, as tree branches that slope upwards. **[164]**

oblong Longer than broad with more or less parallel sides. **[190]**

obovate Inversely ovate, broadest towards the apex and tapering to the stalk. **[190]**

obsolescent Having dwindled to a rudimentary state or vanished altogether.

obtrullate Inversely trullate, i.e. with the two longer sides meeting at the base.

obturator An outgrowth from the placenta over the micropyle that nourishes the pollen tube and guides it towards the ovule. **[160]**

obtuse Blunt. **[194, 195]**

obvolute With half of one leaf wrapped round

果实。 [184, 258]

下垂的 弯曲向下的。 [163, 164]

小坚果 小坚果，尤其指瘦果和分果的分果爿。 [258]

营养，养分 植物生长和发育必需的物质之一（见"大量元素"和"微量元素"）。

感夜性，就眠性（睡眠运动） 植物各部分，特别是花和叶，对夜间黑暗的响应。 [174]

倒锥形的 （见"陀螺状的"）。

倒心形的 最宽处向着微凹的顶端，然后逐渐变窄成柄。 [191]

倒三角形的 具平截顶端，然后逐渐变窄成柄。 [190]

具外轮对瓣雄蕊的 具两轮雄蕊的，内轮对萼片，外轮对花瓣。 [214]

倒披针形的 最宽处向着顶端，然后逐渐变窄成柄。 [190]

扁球形的 （见"近球形的"）。

偏斜的 不等的 像具一侧在另一侧下扩展的叶的；斜升的，像树的分枝倾斜向上生长。 [164]

矩圆形 长大于宽，且各边或多或少平行的几何图形。 [190]

倒卵形的 最宽处向着顶端，然后逐渐变窄成柄。 [190]

消失的，废退的 具减少到不完全发育状态或完全消失的。

倒镘开拓的 倒镘形的，即具两个基部汇合的长边的。

珠孔塞 从胎座上生出、位于珠孔之上、营养花粉管并诱导其到达胚珠的生长物。 [160]

钝的 不锋利的，钝的。 [194, 195]

跨褶的，卷压的 在芽内叶子的一半包住另

half of another leaf in the bud.

occlusion The process by which wounds in trees are healed by the formation of callus.

ochrea A tubular sheath, formed by the fusion of two stipules at the nodes of many plants in the Polygonaceae. **[187]**

ochreate Having ochreae. **[187]**

octamerous Having the parts of the flower in eights.

octandrous Having eight stamens.

octoploid (8n) Having eight sets of chromosomes in each cell.

odd-pinnate (see **imparipinnate**).

odoriferous Having a smell, especially a fragrant one.

odorous Having a distinct smell.

offset, offshoot A short runner producing a new plant at its tip, as in the genus *Sempervivum* (Crassulaceae). **[154]**

oleaceous Relatin g to the Oleaceae.

oleiferous Producing oil.

oligostemonous (paucistemonous) Having few stamens.

ombrophile (see **ombrophyte**).

ombrophobe A plant intolerant of prolonged rainfall.

ombrophyte (ombrophile) A plant adapted to grow in places with prolonged rainfall.

ombrophytic Growing in rainy habitats.

oogamous Involving the union of gametes of dissimilar size.

oogamy The union of gametes of dissimilar size, usually a small motile male gamete and a large non-motile female gamete.

oosphere A female gamete.

oospore The zygote formed from a fertilised oosphere. **[284]**

op. cit. (opere citato) In the work cited, a reference to a publication mentioned earlier in the same account.

open aestivation With the flower parts in the

一半。

阻塞 树木上伤口通过愈伤组织逐渐恢复的过程。

托叶鞘 蓼科许多植物节上的两片托叶愈合形成的筒状鞘。[187]

具托叶鞘的 具有托叶鞘。[187]

八基数的 具以八为基数的花各部分的。

八雄蕊的 具八枚雄蕊的。

八倍体的（8n） 每一细胞具八套染色体的。

奇数羽状的 （见"奇数羽状的"）。

有气味的，有香味的 具气味，特别是芳香性的。

有气味的 有明显气味的。

短匍茎 其顶端产生一新植株的一段短小匍匐茎，如长生草属（景天科）植物。[154]

木犀科的 与木犀科有关的。

含油的 产油。

少雄蕊的 有少数雄蕊的。

喜雨植物，适雨植物 （见"喜雨植物"）。

嫌雨植物 不能容忍长时间下雨的植物。

喜雨植物 适合生长于长时间下雨地区的植物。

喜雨的 在多雨生境生长的。

卵配生殖的 涉及不同大小配子融合的。

卵配生殖 不同大小的配子的融合，通常为小的运动的雄配子和大的不动的雌配子。

卵球 一种雌配子。

合子（尤指卵孢子） 从一个受精的卵球形成的合子。[284]

已引证的文献 指在同一篇文献中前面已经提到的出版物。

张开花被卷迭式 具芽中花各部分不接触的。

bud not touching.

operculate Having an operculum. [256]

operculum The lid of a circumscissile fruit; the membranous cover of the nectar-secreting ring in the genus *Passiflora* (passion flower). [228]

opposite leaves With two leaves at a node, one on each side of the stem or axis. [186]

opposite vernation With the leaves in the bud facing each other but not appressed.

oppositipetalous Situated before a petal.

oppositisepalous Situated before a sepal.

orbicular Circular. [190]

orchid A member of the Orchidaceae. [250—253]

orchidaceous Resembling an orchid flower. [211]

order A taxonomic rank comprising a group of families. Names of orders end in '-ales'.

ornamentation The layer of spines on the tectum of a pollen grain. [212]

ornithophilous Depending on birds to convey pollen for fertilisation.

ornithophily Pollination by birds.

orthotropic Growing directly towards the source of the stimulus (positively orthotropic) or directly away from the source of the stimulus (negatively orthotropic).

orthotropism The growth movement of a plant or part of a plant directly towards or directly away from the source of the stimulus.

orthotropous (atropous) Having the ovule borne on a straight funicle, and with the micropyle in a line with it. [223]

osmosis The movement of molecules from a solution of low concentration to one of higher concentration through a semi-permeable membrane until both solutions are of the same concentration.

ostiole, ostiolum A pore, especially one which acts as an outlet for spores or gametes. [263]

具盖的 有盖。[256]

盖 周裂果实的果盖；西番莲属植物蜜腺环上的膜质盖。[228]

对生叶 茎或轴的节上具两片叶，每侧着生一片叶。[186]

对生叶片卷迭式 芽中具彼此相对但不紧贴的叶的。

对瓣的 位于花瓣前面的。

对萼的 位于萼片前面的。

圆形的 圆形的。[190]

兰花 兰科的植物。[250—253]

兰花的 与兰花有关的。[211]

目 包括一群科的分类单位，目的名称以-ales结尾。

纹饰 花粉粒外壁上的刺层。[212]

鸟媒的 依靠鸟类进行传粉、受精的。

鸟媒 依靠鸟类进行传粉。

直生的 直接趋向刺激物源生长的（正直生的）或直接离开刺激物源生长的（负直生的）。

直生性 植物或植物各部分直接趋向或离开刺激物源的生长运动。

直生的 具着生在直立珠柄上的胚珠，顶端具珠孔。[223]

渗透 一类通过半透膜、从低浓度溶液到高浓度溶液、直到两溶液具有相同浓度前的分子运动。

小孔，孔口 尤其指孢子或配子的出口。[263]

outgrowth A structure growing out from the main body.

oval Broadly elliptical, but ellipsoid when applied to habit. **[163, 173, 190]**

ovary The lower part of a carpel (or carpels) which contains the ovules. **[213, 218, 219, 222]**

ovate With the outline egg-shaped. **[190]**

oviform, ovoid, ovoidal Egg-shaped. **[270]**

ovulate cone A female cone.

ovule A structure which, after fertilisation, develops into a seed. **[213, 219, 222, 223]**

ovuliferous Bearing ovules. **[267]**

ovum (egg) A non-motile female gamete. **[213, 275]**

P (in a floral formula) Perianth, e.g. P6 indicates a perianth composed of 6 perianth segmeats.

pachycaul Thick-stemmed, as palms and cycads, species of *Acropogon* native in New Caledonia, and tree-like herbs with grossly stout stems such as some species of *Lobelia* (giant lobelia) and *Dendrosenecio* (giant groundsel) found in the montane forests of Afried.

pad In the fern genus *Platycerium*, the shield or saddle-shaped sterile fronds at the base of the plant. **[277]**

paired (see **geminate**).

palate The swollen part of the lower lip of a gamopetalous flower which almost or entirely closes the throat, as in the personate flower of *Antirrhinum*. **[211]**

palea The upper of the two bracts enclosing a grass flower. **[243]**

paleaceous Chaffy. **[200]**

palaeobotany The study of fossil plants.

palisade mesophyll The layer(s) of elongated cells that are arranged fence-like immediately beneath and at right-angles to the epidermis

突起，旁枝 一类从主体上生出的结构。

卵形的，卵圆的 阔椭圆形的，但当表示体态时用ellipsoid。 [163, 173, 190]

子房 含有胚珠的心皮下部。 [213, 218, 219, 222]

卵圆形的 具卵形轮廓的。 [190]

卵形的 卵形的。 [270]

近卵球形球果 雌球果。

胚珠 受精后发育成种子的结构。 [213, 219, 222, 223]

具胚球的 生有胚珠的。 [267]

卵，卵子 一类不活动的雌配子。 [213, 275]

花被 （花程式中）花被，例如，P6表示花被由6枚花被裂片组成。

粗茎的 粗茎的，如棕榈类和苏铁类，原产新喀里多尼亚的顶须属植物，和在非洲山地森林中发现的具极粗壮茎的乔木状草本，如半边莲属（巨大的半边莲）和木本千里光属（巨大千里光）的某些植物。

垫 在鹿角蕨属植物中，植物基部的盾状或鞍形不育叶。 [277]

双生的 （见"成对的"）。

喉凸 合瓣花下唇瓣上的膨大部分，它几乎或完全将喉部封住，如金鱼草属植物的假面状花。 [211]

内稃 包在禾本科植物花外的两枚苞片中的上方的一枚。 [243]

稃状的 谷壳状的。 [200]

古植物学 化石植物的研究。

栅栏叶肉 直接位于中生双子叶植物叶表皮之下并与之成直角排列成栅栏状的、伸长细胞层 [196]（又见"海绵组织"）。

in mesophytic dicotyledon leaves. [196] (see also **spongy mesophyll**)

palman Of palms, the undivided central part of a fan palm leaf, e.g. *Chamaerops*. [249]

palmate Divided to the base into separate leaflets, all the leaflets arising from the end of the leaf stalk. [192]; having the veins radiating from the end of the leaf stalk to the tips of the lobes. [189]

palmatifid More or less hand-shaped, with the lobes of the leaf extending about half-way to the base. [191]

palmatipartite More or less hand-shaped, with the lobes of the leaf extending from about half to two-thirds of the way towards the base.

palmatisect More or less hand-shaped, with the lobes of the leaf extending almost to the base. [191]

palus One of the shorter filaments in the corona of plants in the Passifloraceae. [228]

palynology The study of pollen grains and other spores to provide information on the distribution of species in earlier times and to assist in dating geological formations and archaeological remains.

pandurate Fiddle-shaped. [191]

panicle A much-branched inflorescence. [205]

paniculate In the form of a panicle.

pannose Felt-like, composed of densely matted woolly hairs. [201]

papilionaceous Butterfly-like, as the flowers of subfamily Papilionoideae in the Leguminosae, in which the corolla consists of a standard petal that encloses two wing petals in bud, and two lower petals more or less united to form a keel, e.g. *Lathyrus*. [230]

papilla (papula) A small, nipple-shaped projection. [221]

papillate Bearing papillae. [199]

掌心 棕榈类的，扇形棕榈叶的中心不裂的部分，如欧洲矮棕属植物。[249]

掌状的，掌状脉的 分裂至基部成离生的小叶，所有小叶都从叶柄末端生出 [192]；具有从叶柄末端辐射到叶片尖端的叶脉。[189]

掌状半裂的 多少掌状的，叶裂片向基部延伸到中部。[191]

掌状深裂的 多少掌状的，叶裂片向基部延伸到中部至三分之二处。

掌状全裂的 多少掌状的，叶裂片几乎延伸到基部。[191]

丝状体 西番莲科植物副花冠中较短的丝状物之一。[228]

孢粉学 花粉与其他孢子的研究，能够提供早期物种分布信息和协助鉴定地质构造的和考古遗迹的年龄。

提琴形的 提琴状的。[191]

圆锥花序 具许多分枝的花序。[205]

圆锥花序状的 具圆锥花序形式的。

毡毛状的 毛毡状的，由密集缠结的羊毛状的毛组成。[201]

蝶形的 蝶状的，如豆科蝶形花亚科植物花，花冠由一枚旗瓣、两枚芽期被旗瓣所包的翼瓣和下部两枚或多或少合生成龙骨瓣的下花瓣组成，如山黧豆属植物。[230]

乳突 一个小的乳头状突起。[221]

具乳突的 生有乳突的。[199]

papilliform Nipple-shaped.

乳突状的 乳头状的。

papillose (papulose) Covered in papillae.

乳突状的 被乳突的。

pappus The specialised calyx of hairs or scales occurring mainly in the Compositae. **[234]**

冠毛 主要出现于菊科植物的、毛状或鳞片状的特化花萼。 [234]

papula (see **papilla**).

乳突 （见"乳突"）。

papulose (see **papillose**).

具小乳突的 （见"具乳突的"）。

parachute mechanisim A form of seed dispersal (in e.g. *Asclepias*) or fruit dispersal (in e.g. *Taraxacum* and other members of the Compositae) in which a feathery appendage enables the seed or fruit to be carried away by the wind, sometimes for great distances.

伞播机制 种子传播（如马利筋属植物）或果实传播（如蒲公英及菊科其他植物）的方式，其羽毛状附属物能使种子或果实被风吹走，有时到达很远处。

paracladium A portion of an inflorescence, which has the same structure as the whole inflorescence.

副花枝 花序的一部分，其与整个花序具有相似的结构。

parallel With the veins remaining more or less the same distance apart along much of the leaf, a characteristic of the veins in the leaves of the majority of monocotyledons. **[168, 189]**

平行的 叶具大多数保持或多或少相同距离的叶脉的，是大多数单子叶植物叶脉的特征。 [168, 189]

paraphysis (plural **paraphyses**) A sterile filament growing amongst the sporangia of cryptogams. **[280]**

隔丝 一类从隐花植物孢子囊间生出的不育丝。 [280]

parasite A plant that lives on another plant and derives its nourishment from it, e.g. *Cuscuta* (dodder). **[147]**

寄生物 一类生活在另一植物体上、并从其吸取营养物质的植物，如菟丝子属植物（菟丝子）。 [147]

parasitic Growing on another plant and deriving nourishmeat from it.

寄生的 在另一植物体上生长，并从中吸取营养物质的。

parenchyma Succulent tissue, consisting of more or less isodiametric, thin-walled cells, often with intercellular spaces, that is found in e.g. the softer parts of leaves, the pulp of fruits and the pith of stems. **[148, 168]**

薄壁组织 在叶的柔软部分、果肉和茎的髓部发现的、由或多或少等径的薄壁细胞组成的、常具细胞间隙的肉质组织。 [148, 168]

parenchymatous Relating to or consisting of parenchyma.

薄壁组织的 有关或组成薄壁组织的。

parietal placentation The arrangement in which the placentas develop along the fused margins of the carpels of a unilocular ovary. **[222]**

侧膜胎座式 胎座沿单室子房心皮边缘融合线发育的排列方式。 [222]

paripinnate (even-pinnate) Having an equal

偶数羽状的 具同等数目的小叶，缺少顶生

number of leaflets and lacking the terminal leaflet. **[192]**

parted Cut, but not quite to the base.

parthenocarpic Bearing fruits produced without preliminary fertilisation.

parthenocarpy The formation of fruit without preliminary fertilisation and usually without development of seeds.

partial parasite (see **hemi-parasite**)

patent Spreading.

pathogen An organism that causes disease.

paucistemonous (see **oligostemonous**).

pectinate Pinnatifid, with narrow segments set close like the teeth of a comb. **[193, 240]**

pedate Palmately lobed or divided, but with the basal lobes again divided. **[192]**

pedicel The stalk of a single flower. **[184, 229]**

pedicellate Of a flower, stalked.

peduncle The stalk of an inflorescence. **[184, 229]**

pedunculate With a peduncle.

peg root One of the breathing roots that grow upwards from the submerged roots of tropical swamp plants and project upright above the surface of the water or mud. **[147]**

pellucid Translucent, allowing the passage of light.

peloria The abnormal development of a typically zygomorphic flower so that it becomes actinomorphic. **[183]**

peltate Shaped like a disc and attached at the centre of its lower surface to the stalk. **[191, 201]**

pendent, pendulous (pensile) Hanging down. **[164, 251]**

pendulous placentation (see **apical placentation**)

penicillate Streaked, as if with a pencil or brush.

penninerved Having veins that branch pinnately. **[188]**

小叶。 [192]

深裂的 裂开但并未完全裂到基部。

单性结实的 生有不经初期受精而形成的果实。

单性结实 不经初期受精、也无种子产生的果实形成过程。

半寄生物 （见"半寄生物"）。

开展的 开展的。

病原体 一类引起疾病的生物体。

少雄蕊的 （见"少雄蕊的"）。

篦齿状的 羽状半裂，具紧密排列成梳齿状的狭窄裂片。 [193, 240]

鸟足状的 掌状浅裂或全裂，具再次分裂的基部裂片。 [192]

花梗 单朵花的柄。[184, 229]

具花梗的 花具柄的。

花序梗 花序的柄。[184, 229]

具花序梗的 具花序梗的。

桩根 热带沼泽植物从沉水根上向上生长的、伸出水面或淤泥表面之上的直立的呼吸根之一。[147]

透明的 透明允许光通过的。

反常整齐花 典型两侧对称花发生异常发育，使之成为辐射对称花。[183]

盾状的 形似盘状，以其下面中心处附着在柄上。[191, 201]

下垂的 下垂的。[164, 251]

悬垂胎座 （见"顶生胎座"）。

具毛撮的，毛笔状的 具条纹的，像具毛笔或刷子一样。

羽状脉的 具羽状分枝的叶脉。[189]

pensile (see **pendent**).

pentadelphous With five groups of stamens, as some species of *Hypericum*.

pentamerous Having the parts of the flower in fives.

pentandrous Having five stamens.

pentaploid (5n) Having five sets of chromosomes in each cell.

pentarch root A root with five protoxylem strands in the stele. **[149]**

pepo A unilocular, many-seeded, hard-walled berry that forms the fruit of *Cucurbita pepo* (Marrow), *Cucumis melo* (Melon), and some other members of the Cucurbitaceae. **[260]**

perennate To continue to live for a number of years, as a perennial plant.

perennating Surviving from one growing season to another, usually with a dormant (resting) period in between.

perennation The act of living for a number of years.

perennial Living for a number of years; a plant that lives for a number of years.

perfect flower A flower in which both the androecium and the gynoecium are functional.

perfoliate With the bases of two opposite, sessile leaves connate round the stem so that the stem appears to pass through a leaf blade. **[187]**

pergamentaceous Parchment-like.

perianth (perigone) A collective term for the outer, non-reproductive parts of a flower, often differentiated into calyx and corolla.

perianth lobe One of the free parts of the perianth when the lower portions are united into a tube. **[225]**

perianth segment One of the parts of the perianth, often used to describe segments which closely resemble each other, as in the genus *Tulipa* (tulip). **[168, 226]** (see **tepal**)

下垂的 （见"下垂的"）。

五体雄蕊的 具有五束雄蕊的，如金丝桃属的一些植物。

五基数的 具以五为基数的花的各部分。

五雄蕊的 具有五枚雄蕊的。

五倍体（5n） 每一细胞具五套染色体的。

五原型根 一类中柱内具五枚原生木质部脊的根。[149]

瓠果 一类单室、具多数种子和坚硬果皮的浆果，像西葫芦、甜瓜和葫芦科某些其他植物的果实。[260]

多年生的 持续生活多年，像多年生植物。

多年生的 从一生长季生存到另一生长季，通常在期间具休眠期。

多年生 生活多年的行为。

多年生的，多年生植物 生活多年的；生活多年的植物。

完全花 一类雄蕊群和雌蕊群都具功能的花。

茎穿叶的 两枚对生、无柄叶的基部围绕茎合生，致使茎仿佛穿过叶片一样。[187]

羊皮纸状的 羊皮纸状的。

花被 一朵花外围的非生殖部分的集体术语，通常分化为花萼和花冠。

花被裂片 花被下部合生成管而上部的离生部分之一。[225]

花被裂片 花被各部分之一，通常用来描述彼此相似的裂片，像郁金香属植物。[168, 226]（见"花被片"）

perianth tube The tube formed when the lower portions of the perianth segments are united. **[225]**

periblem The intermediate layer of the apical meristem which develops into the cortex.

pericarp The fruit wall that has developed from the ovary wall. **[158]**

pericycle The outermost layer of cells of the stele in a stem or root. **[149, 169, 268]**

periderm Secondary protective tissue, consisting of phellem (cork), phellogen (cork cambium), and phelloderm, that often replaces the epidermis in older stems and roots. **[171, 173]**

perigone (see **perianth**).

perigynium The utricle which encloses the female flower in some members of the Cyperaceae. **[246]**

perigynous With the sepals, petals and stamens inserted around the ovary on the hypanthium, a concave structure developed from the receptacle. **[218]**

perisperm The albumen of a seed formed outside the embryo sac.

perpetual Flowering several times in a season, as some rose cultivars.

persistent Remaining attached.

personate Bilabiate with a prominent palate. **[211]**

perular scale One of the basal scales of leaf buds which may persist for some time after the development of the shoot.

petal A single segment of the corolla. **[232]**

petaloid Petal-like, as the style in the genus *Iris* [184, 211, 225]; a petal-like structure bearing distorted anthers, situated between the normal petals and the stamens in semi-double and double flowers.

petiolar Relating to a petiole.

petiolate Having a leaf stalk. **[187]**

petiole A leaf stalk. **[151, 177—180]**

花被筒 花被裂片下部合生形成的筒。[225]

表皮原 顶端分生组织中发育成皮层的过渡层。

果皮 由子房壁发育而来的果壁。[158]

中柱鞘 茎或根中柱的最外层细胞。[149, 167, 268]

周皮 由木栓层、木栓形成层和栓内层组成、通常在老茎和老根上代替表皮的次生保护组织。[171, 173]

花被 （见"花被"）。

雌器苞（果囊） 莎草科某些植物上包住雌花的卵形囊。[246]

周位的 萼片、花瓣和雄蕊围绕子房着生于花筒边缘的，花筒是由花托发育而来的杯状结构。[218]

外胚乳 在胚囊外形成的种子胚乳。

连续开花的 一季节内数次开花的，像某些玫瑰品种。

宿存的 保持附着的。

假面状的 二唇形具一突出喉凸（下唇瓣）的。[211]

叶芽鳞片 叶芽的基生鳞片之一，可在芽发育后继续存在一段时间。

花瓣 花冠的单个裂片。[232]

花瓣状的 花瓣状的，像鸢尾属植物的花柱 [184, 211, 225]；一类生有扭曲花药、位于正常花瓣与半重瓣花和重瓣花的雄蕊之间的花瓣状结构。

叶柄的 与叶柄有关的。

具叶柄的 具有叶柄的。[187]

叶柄 叶柄。[151, 177—180]

petiolule The stalk of a leaflet in a compound leaf.

petrophilous Adapted to growing in a rocky environment.

pH A value of hydrogen-ion concentration which allows the acidity or alkalinity of a solution to be measured on a scale from 1 (extremely acid) to 15 (extremely alkaline). (see **acid soils**, **neutral soils**, and **alkaline soils**)

phanerogam A plant belonging to the Phanerogamia, a division in former classifications that included all the seed-bearing plants, now called spermatophytes.

phanerophyte A tall, woody or herbaceous perennial, with resting buds more than 25 cm above soil level. [166]

phellem (cork) A spongy, protective layer of thin-walled cells impregnated with suberin that is formed from cork cambium, and often replaces the epidermis in stems and roots as they grow older. [171]

phelloderm One or more layers of thin-walled cells formed from the inner side of the phellogen. [171]

phellogen (cork cambium) The layer of meristematic cells lying just beneath the surface of a stem or root, that forms phellem (cork) on its outer side and phelloderm on its inner side. [171]

phenetic classification A type of classification which expresses relationships between plants in terms of their visible or otherwise measurable physical and biochemical characteristics.

phloem (bast) The vascular tissue that conducts sap containing nutrients produced by photosynthesis from the leaves to other parts of the plant. [171]

photonasty The response of a plant to a change in light intensity, e.g. the opening

小叶柄 在复叶上的小叶的柄。

岩生的 适于岩石环境生长的。

酸碱度 氢离子浓度的负对数，允许一种溶液的酸碱度用一个从1（极酸性）到15（极碱性）的等级来测量。（见"酸性土壤""中性土壤"和"碱性土壤"）。

显花植物 一类属于显花植物门的植物，在前期分类中，该门包括所有生有种子的植物，现在被称为种子植物。

高位芽植物 一类高大的木本或草本的多年生植物，其休眠芽位于土壤面之上25厘米以上。[166]

木栓 由木栓形成层形成的以软木脂灌注的薄壁细胞构成的海绵状保护层，在茎和根中当表皮衰老时通常替代它。[171]

栓内层 从木栓形成层内侧形成的一至多层薄壁细胞。[171]

木栓形成层 位于茎或根表皮下的分生组织细胞层，在其外面形成木栓，在其内面形成栓内层。[171]

表征分类 一类根据可见或其他可测量的生理生化性状表达植物之间关系的分类。

韧皮部 将光合作用产生的含营养的液体从叶输导到植物体各部分的维管组织。[171]

感光性 植物对光强度改变的响应，如花的开放和闭合。

and closing of the flowers.

photoperiodism The response of a plant to the relative duration of day and night, especially in regard to flowering.

photosynthesis The process by which green plants convert carbon dioxide and water into carbohydrates in the presence of sunlight.

photosynthetic Relating to photosynthesis. **[147]**

phototropic Turning towards the light source (positively phototropic) or away from it (negatively phototropic).

phototropism The growth movement of plants in response to the stimulus of light. **[173]**

phyllary One of the involucral bracts that surround the head of flowers in the Compositae. **[234]**

phylloclade (see **cladode**).

phyllode A petiole taking on the form and functions of a leaf, as in the genus *Acacia*. **[177]**

phyllopodium An outgrowth on the rhizome of some ferns on which a frond arises. **[278]**

phyllotaxis, phyllotaxy The arrangement of leaves on an axis or stem. **[175]**

phylogenetic classification A type of classification which expresses supposed relationships between plants in terms of their evolutionary history.

phylogeny The relationships between plants as determined by their evolutionary history.

phylum (see **division**).

phytochrome A light-sensitive pigment in plants that is involved with certain developmental processes such as photoperiodism, reversal of etiolation, and the germination of some seeds and spores.

phytomer A bud-bearing node, the smallest structural unit of a plant which is capable of reproducing vegetatively.

光周期现象 植物对白天和黑夜相对持续时间的响应，特别对于开花。

光合作用 绿色植物在阳光存在时将二氧化碳和水转化成碳水化合物的过程。

光合的 与光合作用有关的。[147]

向光性的 朝向光源（正向光性）或离开它（负向光性）。

向光性 对光刺激做出响应的生长运动。[173]

总苞片 围绕在菊科植物头状花序周围的总苞片之一。[234]

叶状枝 （见"叶状枝"）。
叶状柄 具叶形态和功能的叶柄，如金合欢属植物。[177]

叶足 某些蕨类植物根状茎上的生长物，其上可生叶。[278]
叶序 叶在轴或茎上的排列方式。[175]

系谱分类 反应根据植物进化历史确定的植物间假定关系的一种分类。

系统发育 由植物进化历史确定的植物之间的关系。
门 （见"门"）。
光敏色素 植物体内涉及某些发育过程如光周期、黄化逆转和某些种子和孢子萌发的感光色素。

植物繁殖单位（如植物茎上的节等） 一类生有芽的节，是植物进行营养繁殖的最小结构单位。

pigment The natural colouring matter in plant tissues. (see **chloroplast** and **chromoplast**)

piliferous layer (epiblem, epiblema, rhizodermis) The epidermis of a young root that bears the root hairs. **[148, 149]**

pilose Softly hairy. **[200]**

pinetum A collection of conifers, especially *Pinus* (pine) species and varieties, for scientific or ornamental purposes.

pin-eyed One of the two forms of a dimorphic flower, e.g. *Primula vulgaris* (Primrose), where the style is long and the stamens are below the stigma. **[229]** (see also **thrum-eyed**)

pinna A primary division or leaflet of a compound leaf. **[192]**

pinnate Having separate leaflets along each side of a common stalk **[192]**; having separate veins along each side of the midrib of a leaf. **[189]**

pinnatifid Pinnately lobed, the lobes extending from about a quarter to half-way towards the rachis. **[191]**

pinnatipartite Pinnately divided, the divisions extending from about half to two-thirds of the way towards the rachis. **[191]**

pinnatisect Pinnately digided, the divisions extending almost to the rachis. **[191]**

pinnule A secondary division of a pinnate leaf. **[275]**

pinnulet A segment of a pinnule. **[276]**

Pinopsida (Coniferae) The class that comprises the conifers. (see **Gymnospermae**)

pistil A single carpel in an apocarpous flower, or the gynoecium in a syncarpous flower. **[219]**

pistillate (carpellate) Having only female organs.

pistillode A sterile pistil.

pitcher plant A carnivorous plant that obtains its nutrients from insects lured into traps

色素 植物组织内的自然有色物质（见"叶绿体"和"有色体"）。

根毛层（根被皮） 生有根毛的幼根表皮。[148, 149]

具柔毛的 具柔软的毛。[200]

松柏园，基于科学或观赏用途而建立的关于球果类尤其是松属植物种与变种的收集园地。

柱头上位的，柱头可见的（针式型的） 如报春花两型花的两种形式之一，其花柱长，雄蕊花药在柱头之下。[229]〔又见"雄蕊上位的（线式型的）"〕

羽片 复叶的初级裂片或小叶。[192]

羽状的 沿着公共柄每侧具离生小叶的[192]；沿着叶中脉每侧具离生叶脉的。[189]

羽状半裂的 羽状裂，裂片向中脉延伸到四分之一至一半处。[191]

羽状深裂的 羽状裂，裂片向中脉延伸到一半至三分之二处。[191]

羽状全裂的 羽状裂，裂片几乎延伸到中脉。[191]

小羽片 羽状叶的次级裂片。[275]

末回裂片 小羽片的裂片。[276]

松柏纲（球果纲） 包括球果类的一类植物（见"裸子植物门"）。

雌蕊 离生心皮花的单个心皮或合生心皮花的雌蕊群。[219]

雌蕊的（心皮的） 仅具雌性器官。

退化雌蕊 不育雌蕊。

捕虫植物 一类从被引诱到囊状叶形成的陷阱的昆虫来获取自身营养的肉食植物，

formed by its pitcher-shaped leaves, e.g. *Sarracenia*, *Darlingtonia*, *Nepenthes*, etc. **[181]**

如瓶子草属、眼镜蛇草属、猪笼草等。[181]

pith The central column of spongy, parenchymatous tissue in the stems of dicotyledons and certain monocotyledons such as *Juncus* (rush); the medulla. [149, 169]

髓 双子叶植物和某些单子叶植物如灯芯草茎中，由海绵状薄壁组织构成的中心柱；髓部。[149, 169]

pith ray (see **medullary ray**).

髓射线 （见"髓射线"）。

pitted With small depressions on the surface.

具洼点的 表面具小凹陷的。

placenta The part of the ovary to which the ovules are attached. **[213, 222]**

胎座 子房内胚珠着生的部位。[213, 222]

placental Relating to a placenta.

胎座的 与胎座相关的。

placentation The arrangement of placentas in an ovary. **[222, 223]**

胎座 子房内胎座的排列方式。[222, 223]

plagiotropic Growing at an angle towards or away from the source of the stimulus.

斜向（性）的 以一定角度朝向或离开刺激物源方向生长的。

plagiotropism The growth movement of a plant or part of a plant at an angle to the source of the stimulus.

斜生性 植物或植物各部分以一定角度朝向或离开刺激物源方向的生长运动。

plantlet A small plant, as those formed on the leaves of some species of flowering plants, e.g. *Kalanchoe*, or on the fronds of some ferns. **[155]**

小植株 有花植物某些种类从叶上形成的小植株，如长寿花或某些蕨类植物的叶。[155]

plastid A specialised structure in the cytoplasm of a plant cell. (see **chloroplast, chromoplast** and **leucoplast**)

质体 植物细胞细胞质中的一种特化结构（见"叶绿体"、"有色体"和"白色体"）。

plectostele A type of protostele, in which the xylem and phloem form more or less parallel bands.

编织中柱 一类木质部和韧皮部形成或多或少平行带的中柱。

pleiochasium A cymose inflorescence in which the main axis has more than two lateral branches. **[206]**

多歧聚伞花序 一类中轴上具两个以上侧枝的聚伞花序。[206]

plerome The innermost layer of an apical meristem which develops into the central vascular tissue.

中柱原 发育成中央维管组织的顶端分生组织最内层。

plica A fold.

皱襞，褶 折叠。

plicate Folded more than once lengthwise. **[188]**

折扇状的 多于一次纵向折叠的。[188]

plumose Feathery. **[147, 221]**

羽状的 羽毛状的。[147, 221]

plumule The young shoot as it emerges from the seed on germination, usually after the

胚芽 一类在胚根出现后、从萌发种子上生出的幼芽。[157, 159, 174]

appearance of the radicle. **[157, 159, 174]**

pluricarpellate Having several carpels.

plurilocular Composed of several or many loculi or compartments.

pneumatophore One of the breathing roots that grow upwards from the submerged roots of mangroves and other tropical swamp plants. They project above the surface of the water or mud as peg roots or knee roots, and have lenticels in their bark that allow air to pass through into the root system. **[147]**

pocket nectary A type of nectary found in e.g. *Ranunculus repens*, in which nectar is secreted beneath a small scale at the base of a petal. **[231]**

pod A dry, many-seeded, dehiscent fruit, particularly the legume in plants of the Leguminosae. **[255]**

podium A base or supporting structure, e.g. the stalk bearing the strobili in *Lycopodium*. **[282]**

poikilohydrous Of a plant, having its water content determined by the wetness or dryness of its surroundings.

polar nuclei The two nuclei that move from the poles of the embryo sac to its centre. **[213]**

pole One of two opposite areas of a more or less spherical structure, e.g. one of the two opposite areas of a pollen grain that are free from apertures. **[212]**

pollen The small grains which contain the male reproductive cells of the flower. **[212]**

pollen flower A flower without nectar that attracts insects by its pollen.

pollen sac One of the two portions into which a theca or anther cell is divided.

pollen tube An outgrowth from a germinating pollen grain which carries the male gamete(s) down the style to an ovule in the

多心皮的 具数枚心皮的。

多室的 由数个或多个室或腔组成的。

出水通气根 从红树和其他热带沼泽植物沉水根上向上生长形成的呼吸根之一。它们以桩根或膝根形式突出水面或淤泥表面以上，其树皮上有允许空气通过并进入根系的皮孔。 [147]

囊状蜜腺 一种在匍枝毛茛中发现的、位于花瓣基部小鳞片下的蜜腺。 [231]

豆荚 一类干燥、具多枚种子、开裂的果实，特别是豆科植物的荚果。 [255]

柄 一种基部的支持结构，如石松属植物着生孢子叶穗的柄。 [282]

变水的（适应各种水质的） 一类其体内含水量由所在环境的干湿度决定的植物。

极核 从胚囊极点移到中心的两个核。 [213]

极点 一类近球形结构的两个相对区域之一，如无萌发孔花粉粒的两个相对区域之一。[212]

花粉 花上含有雄性生殖细胞的小颗粒。 [212]

产粉花 一类不具有蜜汁靠花粉吸引昆虫的花。

花粉囊 花粉囊或药室被分隔成的两部分之一。

花粉管 正在萌发的花粉粒上的管状生长物，能够运送雄配子向下经花柱进入子房内的胚珠。 [213]

ovary. **[213]**

pollinarium A collective term for the pollinia, stipe, and viscidium in flowers of the Orchidaceae. **[252]**

pollination The placing of pollen on the stigma or stigmatic surface. **[213]**

polliniferous Bearing pollen.

pollinium A pollen mass, as in Orchidaceae and Asclepiadaceae. **[233, 252]**

polyad A group of more than four pollen grains.

polyadelphous With more than two groups of stamens, as in some species of the genus *Hypericum*. **[217]**

polyandrous (polystemonous) Having numerous stamens.

polyarch root A root with many protoxylem strands in the stele. **[149]**

polygamo-dioecious Functionally dioecious, but bearing a few flowers of the opposite sex or a few perfect flowers on the same plant.

polygamo-monoecious Polygamous, but in the main monoecious.

polygamous Bearing both unisexual and bisexual flowers on the same plant.

polygynous Having many styles.

polymerous With numerous members in each series or whorl.

polymorphic Having several or many forms.

polymorphism The occurrence of different forms of a plant species within a particular population.

polypetalous (see **apopetalous**).

polyphyllous Having separate petals and sepals.

polyploid (many-n) Having more than the usual two sets of chromosomes in each cell.

polysepalous (see **aposepalous**).

polystemonous (see **polyandrous**).

polystichous Arranged in several or many

花粉器 兰科植物花的花粉块、柄和粘盘的集合名词。 **[252]**

传粉 将花粉放到柱头或柱头表面上。 **[213]**

具花粉的 生有花粉的 。

花粉块 一类花粉团，常见于兰科和萝藦科植物。 **[233, 252]**

多合花粉 多于四个花粉粒的组合。

多体雄蕊的 具两束以上的雄蕊，如金丝桃属某些植物。 **[217]**

多雄蕊的 具多数雄蕊的。

多原型根 一类中柱中具许多原生木质部脊的根。 **[149]**

杂性异株的 功能性雌雄异株的，但在同一植株上还具少数相异单性花和少数完全花。

杂性同株的 杂性的，但以雌雄同株为主。

杂性的 单性花和两性花生于同一植株上的。

多雌蕊的 具有许多花柱的。

多基数的 一轮中具多个部分的。

多态的 有数种或多种形式的。

多态性 在一特定居群内一植物物种出现不同变型的现象。

离瓣的 （见"离生花瓣的"）。

离生花被的（多叶的） 具离生花瓣或萼片的。

多倍体（many-n） 每一细胞内具两套以上染色体的。

离萼的 （见"离生萼片的"）。

多雄蕊的 （见"多雄蕊的"）。

多列的 排成几列或多列的。

rows.

polytelic An inflorescence in which the branches do not end in a flower.

pome A fruit consisting of a core, formed by several united carpels, enclosed within a firm, fleshy receptacle, as in the genus *Malus* (apple), and other members of subfamily Maloideae in the Rosaceae. **[261]**

pomology The science and practice of fruit culture.

pore A small, usually round aperture. **[212, 256]**

poricidal, porose Opening by pores, as the anthers in many members of the Ericaceae and the Polygalaceae, or the capsules in some genera, e.g. *Papaver* (poppy). **[216, 256]**

porogamy The entry of the pollen tube into the ovule through the micropyle.

posterior Back, towards the axis. **[226]**

postichous On the posterior side, next the axis.

pouch A bag-shaped structure.

p. p. (pro parte) In part, added to a plant name to indicate that the name now only partially covers the former concept of the taxon.

praemorse Irregularly truncate, appearing as if bitten off at the apex. **[195]**

precocious flower A flower that opens early in the season, before the leaves appear.

prickle A sharp-pointed outgrowth from the superficial tissues of the stem, as in the genus *Rosa*. **[180]**

primary meristem Tissue derived from the apical meristem, and appropriate to the part concerned.

primary phloem Phloem tissue derived from the procambium during the growth of a vascular plant.

primary xylem Xylem tissue derived from

多顶枝的 一类分枝末端不是花的花序。

梨果 一类具一枚由多枚合生心皮发育形成的、包在坚实肉质花托内的果核的果实，像苹果属植物（苹果）和其他蔷薇科苹果亚科植物。[261]

果树学 果树栽培的科学和实践。

孔 一个小的圆形孔。[212, 256]

孔裂的 以孔开裂的，如杜鹃花科和远志科许多植物的花药，或某些属如罂粟属植物（罂粟）的蒴果。[216, 256]

珠孔受精 花粉管通过珠孔进入胚珠。

后面的 后面的，朝向轴。[226]

后的，近轴的 在后面的，近轴的。

囊，袋 一种袋状结构。

一部分，部分地 用在植物名称之后，表示该名称现在仅部分涵盖类群的原先概念。

啮蚀状的 不规则平截的，似乎在顶端被咬掉了。[195]

先叶花 花在叶露出之前就早已开放。

皮刺 从茎的表面组织生出的尖锐突起物，像蔷薇属植物。[180]

初生分生组织 来源于顶端分生组织，并适合于相关部位的组织。

初生韧皮部 维管植物生长过程中，从原形成层发育而来的韧皮部组织。

初生木质部 维管植物生长过程中，从原形

the procambium during the growth of a vascular plant.

procambium The cells of primary meristem which differentiate into primary xylem, primary phloem, and cambium.

procumbent Lying along the ground. **[162]**

progeny Offspring, immediate descendents.

prolepsis Growth of a bud into a lateral shoot after a period of dormancy.

proleptic Growing into a lateral shoot from a dormant bud.

proliferous Bearing gemmae or plantlets. **[155]**

proliferating Producing buds in the axils of perianth segments.

proliferation In some angiosperms the development of buds in the axils of perianth segments, or in some gymnosperms the development of a leafy shoot from a female cone.

prolification (see **proliferation**).

prone (see **Prostrate**).

propagation The multiplication of plants by seeds or various kinds of vegetative material, either under natural conditions or under the sterile conditions of *in vitro* culture.

propagule Any structure capable of giving rise to a new plant by sexual or asexual means. **[155]**

prophyll The lowest of the papery structures attached to the scape in the genus *Crocus* (Iridaceae). **[150]**

prop root (stilt root) An adventitious root that grows out from the lower part of a stem into the soil to support that stem, or grows down from a lower branch into the soil to support that branch. **[146, 149]**

prostrate (prone) Lying flat. **[164]**

pro syn. (pro synonymo) A phrase used after a plant name to indicate that it was first published as a synonym.

成层发育而来的木质部组织。

原形成层 可分化为初生木质部、初生韧皮部和形成层的初生分生组织细胞。

平卧的 俯卧在地面上的。[162]

后裔 后代，直系后裔。

预兆 芽休眠期后生长成侧枝。

预兆的 从休眠芽长成侧枝的。

多育的 生有芽或小植株的。[155]

层出的 花被片腋内产生芽。

层出（现象） 某些被子植物花被片腋内芽的发育，或某些裸子植物从雌球花生出带叶枝条的发育。

层出（现象） ［见"层出（现象）"］。

前倾的 （见"平卧的"）。

繁殖 无论在自然条件还是在实验室的无菌条件下，植物通过种子或各种营养材料的增殖。

繁殖体 通过有性或无性方式能够生出新植株的任一结构。[155]

先出叶 番红花属（鸢尾科）植物着生于花葶上的最下方的纸质结构。[150]

支柱根 从茎下部生出进入土壤以支持该茎，或从下方枝条向下生长进入土壤以支持该枝的一种不定根。[146, 149]

平卧的 平卧的。[164]

作为异名 一个用在植物名称后的短语，表示该名称是首次作为异名发表。

protandrous Having stamens which mature and shed their pollen before the stigmas of the same flower become receptive.

protandry The state of being protandrous.

prothallial Relating to the prothallus in ferns and fern allies; the kind of cells that occur within the wall of the pollen grain of e.g. *Pinus* (pine). **[267]**

prothallus The plantlet that develops from the spore of a fern or other vascular cryptogam. It lacks true roots, stems, and leaves, but bears male and/or female sex organs, and is attached to the soil by rhizoids **[275, 282—284]**; the group of prothallial cells in the ovule of a gymnosperm.

protogynous Having stigmas which become receptive before the stamens of the sarne flower mature and shed their pollen.

protogyny The state of being protogynous.

protophloem The first phloem that is formed from the procambium.

protoplasm The living material within a plant cell, consisting of the nucleus and plastids embedded in cytoplasm.

protostele A type of stele that lacks pith, and consists only of xylem and phloem. **[280]** (see **haplostele**, **actinostele**, and **plectostele**)

protoxylem The first xylem that is formed from the procambium.

provernal Occurring in early spring.

proximal Situated near to the point of attachment.

pruinose Having a bloom.

pseudanthium A group of small flowers that collectively simulate a single flower, as *Cornus florida*, where the flowers are surrounded by large petaloid bracts. **[208]**

pseudobulb A bulb-like enlargement of the stem in orchids. **[250]**

pseudobulbous Having a pseudobulb.

雄蕊先熟的 雄蕊在同一花的柱头能接受花粉之前已经成熟并散出花粉。

雄蕊先熟 一种雄蕊先熟的状态。

原叶体的 与蕨类植物原叶体有关的；一类发生在花粉粒壁上的细胞，如松。 [267]

原叶体 由蕨类或其他维管隐花植物的孢子发育而来的小植株。它缺少真正的根、茎和叶，但生有雌性和（或）雄性的性器官，并以假根着生于土壤上 [275, 282—284]；一群位于裸子植物胚珠内的原叶体细胞。

雌蕊先熟的 柱头在同一花的雄蕊成熟并散出花粉之前就已经能够接受花粉。

雌蕊先熟 一种雌蕊先熟的状态。

原生韧皮部 由原形成层发育而来的最初的韧皮部。

原生质 植物细胞内的生命物质，包括细胞核、细胞质及其内含的质体。

原生中柱 一种缺乏髓的中柱，只包括木质部和韧皮部 [280]（见"单中柱"、"星状中柱"和"编织中柱"）。

原生木质部 由原形成层发育而来的最初的木质部。

早春的，早春开花的 早春出现的。

近基的 位于着生点附近的。

具粉霜的 具粉霜的。

假单花花序 集合在一起模拟单朵花的一群小花，如北美山茱萸，其花被大的花瓣状苞片所包围。 [208]

假鳞茎 兰花茎的鳞茎状膨大物。 [250]

假鳞茎的 具假鳞茎的。

pseudocarp (false fruit) A structure comprising the mature ovary combined with some other part of the plant, as the 'hip' in the genus *Rosa*. **[261]**

假果 由成熟子房结合某些其他部分构成的结构，如蔷薇属植物的"蔷薇果"。[261]

pseudocephalium The dense mass of hair at the top of the stem in certain cacti.

假花座 某些仙人掌茎顶端的密集毛簇。

pseudostipule One of the lowermost leaflets in the compound leaves of some dicotyledons and therefore very close to the point of insertion of the leaf and the true stipules if these are present.

假托叶 某些双子叶植物复叶最下方的小叶之一，因此若叶和真正托叶存在，则该小叶非常接近其着生点。

pseudoterminal Of a bud or flower, giving the appearance of being at the apex of the stem, but in fact axillary.

假顶生的 指芽或花，在外观上看生在茎顶端，但实际生在叶腋内。

Pteridophyta, pteridophytes Ferns and fern allies. **[275—284]**

蕨类植物门，蕨类植物 蕨类和拟蕨类。[275—284]

ptyxis The way in which an individual leaf is folded within a vegetative bud. **[188]** (see also **vernation**)

个叶卷叠式 一枚叶芽内单叶折叠的方式。[188]（又见"多叶卷叠式"）

puberulent Minutely Pubescent.

被微柔毛的 具小柔毛的。

puberulous Slightly hairy. **[200]**

被微柔毛的 稍有微柔毛的。[200]

pubescence Hairiness.

短柔毛 短柔毛。

pubescent (downy) Covered in soft hairs. **[201]**

被柔毛的 覆盖以柔毛的。[201]

pullulate To germinate, bud, or sprout.

（种子）发芽的 萌发，出芽或嫩梢。

pulse The edible seeds of plants in the Leguminosae cultivated as food crops, such as peas, beans, lentils, etc.

豆类 豆科植物中被栽培用作粮食作物的可食用种子，像豌豆、蚕豆、小扁豆等。

pulvinate Cushion-like.

垫状的 垫子状的。

pulvinus An enlarged portion of the petiole, at its base in *Mimosa pudica* (Sensitive Plant), a member of the Leguminosae **[180]**, or at its junction with the lamina in members of the Marantaceae. **[178]**

叶枕 豆科植物含羞草的叶柄基部 [180]、或竹芋科植物叶柄与叶片相接处的叶柄膨大部分。[178]

punctate Marked with dots, depressions, or translucent glands. **[199]**

具点的 有点、凹陷、半透明腺体。[199]

punctiform In the form of a dot or point.

点状的 以点的形式。

punctum A dot-like marking on any plant organ.

斑点 任一植物器官上的点状斑纹。

pungent Ending in a stiff, sharp point. **[195]**

锐尖的 终止于一坚硬、尖锐的点。[195]

pusticulate Bearing minute, pimple-like

具小疱状突起的 生有小的、疱状突起的。

protuberances.

pustulate Covered with pustules. **[199]**

pustule A pimple-like projection from the surface.

pyramidal Conical in habit. **[163, 164]**

pyrene The stone of a drupe, consisting of the seed surrounded by the hard endocarp. **[260]**

pyriform Pear-shaped.

pyxidium (see **pyxis**).

pyxis (plural **pyxides**) A capsule with circumscissile dehiscence, as in the genus *Anagallis* (Primulaceae). **[256]**

quadrifoliate Four-leaved.

quadrifoliolate With four leaflets. **[281]**

quadrigeneric Composed of four different genera.

quadrilocular Having four loculi or compartments. **[219]**

quadripartite Divided almost to the base into four parts.

quincuncial Partially imbricated of five parts, two being exterior, two interior, and the fifth having one margin exterior and the other interior, as the sepals in the genus *Rosa*.

quinquelocular Having five loculi or compartments.

race A group of individuals within a species that show some similarities but are not sufficiently distinct to constitute a separate species.

raceme An indeterminate inflorescence with pedicillate flowers. **[204]**

racemose In the form of a raceme, bearing racemes. **[184]**

rachilla A secondary axis in the inflorescence of grasses. **[243]**

rachis (plural **rachides or rachises**) The axis of a compound leaf or an inflorescence. **[272, 275]**

radial Extending from a common centre. **[240]**

多疱状突起的 被疱状突起的。 [199]

小疱状突起 从表面生出的疱状突起物。

金字塔形的 （树木）体态呈宝塔状的。 [163, 164]

分核，核果的坚硬部分，由种子和包围在其外的坚硬内果皮组成。 [260]

梨形的 梨形的。

盖果 （见"盖果"）。

盖果（复数为pyxides） 周裂的蒴果，如海绿属植物的果实（报春花科）。[256]

四叶的 具四枚叶的。

四小叶的 具四枚小叶的。 [281]

四属的（尤指杂种） 由四个不同的属组成的。

四室的 具四个腔或室的。 [219]

四深裂的 几乎裂至基部成四部分。

双盖覆瓦状的 五部分覆瓦状排列的，两片在外，两片在内，第五片一个边缘在外，一个边缘在内，像蔷薇属植物的萼片。

五室的 具五个腔或室。

种族 一类种内虽显示出某些相似性、但无足够区别构成一独立种的个体群。

总状花序 一类由具柄花构成的无限花序。 [204]

总状花序状的，生有总状花序的。 [184]

小穗轴 禾本科植物花序的次生轴。 [243]

轴（复数为rachides或rachises） 复叶或花序的轴。 [272, 275]

辐射状的 从一共同中心扩散。 [240]

radially symmetric (see **actinomorphic**).

radiate Spreading from a common centre [139]; having a capitulum with ray florets, as in some members of the Compositae. [221]

radical Of leaves, arising directly from the rootstock. [245]

radicant Rooting.

radicel A small root, a rootlet.

radicle The young root as it emerges from the seed, normally the first organ to appear on germination. [157, 158, 174]

radius One of the longer filaments in the corona of plants in the Passifloraceae. [228]

rambler Any of the cultivated varieties of *Rosa* (rose) which straggle over adjacent vegetation, fences, walls, etc.

ramentum One of the thin, dry scales that often occur on the stems and leaves of ferns. [277]

ramet An individual member of a clone.

ramiflory The production of flowers on the branches of trees (included in the term cauliflory).

ramose Branching, having many branches.

rank A vertical row of lateral organs; the name of a taxonomic unit, e.g. genus, species, etc.

raphe The longitudinal ridge that represents the part of the funicle that is fused to the ovule or seed.

raphide One of the needle-shaped crystals, usually of calcium oxalate, found in the cells of some plants.

Raunkiaer A Danish ecologist (Christen Raunkiaer, 1860—1938) who devised a classification of plants based on the position of their perennating buds in relation to the soil surface, and a classification of leaves according to their surface area. [166]

ray A primary branch of an umbel [205]; the

辐射对称的 （见辐射对称的）。

辐射状的 从一共同中心展开；具有舌状花的头状花序，如菊科的一些植物。[221]

根生的 直接从根茎上生出叶的。[245]

生根的 生根的。

小根 小根。

胚根 从种子露出的幼根，通常为种子萌发时出现的第一器官。[157, 158, 174]

辐射丝 西番莲科植物副花冠的长丝之一。[228]

蔓生植物 蔓延在邻近植被、篱笆、墙等上的任一蔷薇品种（玫瑰）。

小鳞片 蕨类植物茎或叶上出现的薄干燥鳞片之一。[277]

无性系分株 无性系分株之一。

枝花，枝花现象 在树枝上生花的现象（包括在术语"（老）茎生花现象"中）。

多枝的 具有多数分枝的。

列，等级 一纵行的侧生器官；分类单位的名称，如属、种等。

种脊 代表和胚珠或种子愈合的珠柄部分的纵脊。

针晶体 在某些植物细胞中发现的、通常为针状的草酸钙晶体之一。

瑙基耶尔 丹麦生态学家（Christen Raunkiaer, 1860—1938）。基于与土壤表面相关的植物休眠芽的位置提出了一植物分类系统；根据叶表面面积提出了一叶分类系统。[166]

伞辐，舌状花冠 伞形花序的初级分枝

extended, strap-like portion of a ray floret in the Compositae. **[234]**

ray floret One of the outer, irregular flowers in the flower heads of some plants in the Compositae. **[234]**

raylet A smaller, secondary branch of an umbel. **[205]**

reaction wood Wood of distinctive structure formed by trees in branches and inclined trunks in order to maintain them in the appropriate position. (see **compression wood** and **tension wood**)

receptacle (thalamus, torus) The end of the stem which bears the flower parts. **[230, 261]**

receptacular Relating to the receptacle. **[234]**

recurved Curved backwards.

reduplicate Folded outwards or downwards. **[248]**

reflexed Bent abruptly backwards. **[229, 270]**

regma (plural **regmata**) A type of schizocarp in which the mericarps dehisce by elastic movement to release their seeds, as in *Geranium*.

regular (see **actinomorphic**).

relict A plant that has survived from a previous age or, because of changed circumstances, is growing in a place geographically remote from its most closely related species.

reniform Kidney-shaped. **[191, 262]**

repand With a slightly sinuate margin. **[193]**

repent (reptant) Creeping along the ground. **[162]**

replicate Explicative, with the leaf margins in the bud folded back, as in some species of the genus *Galanthus* (snowdrop); of a cell, to reproduce itself.

replum (false septum) A partition formed by outgrowths of the placentas, as in fruits of the Cruciferae. **[264]**

[205]；菊科植物舌状花的伸展的带状部分。 [234]

舌状花，边花 菊科某些植物头状花序外轮的不规则花之一。 [234]

小伞辐 伞形花序的更小的、次级分枝。 [205]

应力木 （见"应力木"和"应拉木"）。

花托 着生花各部分的茎或枝的顶端。 [230, 261]

花托的 与花托相关的。 [234]

后弯的 向后弯曲的。

外向镊合状 向外或向下折叠的。 [248]

反折的 向后突然弯曲的。 [229, 270]

弹裂蒴果（复数为regmata） 一种类型的分果，分果爿通过弹力运动开裂以释放其种子，像牻牛儿苗属植物。

整齐的 （见"辐射对称"）。

孑遗植物 从以前地质年代生存下来的植物，由于环境的改变，生长于远离它最近相关种类的地理区域。

肾形的 肾形的。 [191, 262]

浅波状的 具有稍波状边缘的。 [193]

匍匐的 沿地面匍匐的。 [162]

反折的，复制 像雪片莲属某些种具芽中向后折叠的叶边缘的；能自我繁殖的细胞的。

假隔膜 通过胎座生长物形成的隔膜，像十字花科植物的果实。 [264]

reproduction The production of new individuals by existing ones.

reptant (see **repent**).

resin A sticky, sometimes fragrant substance, insoluble in water, which is secreted by special cells in conifers for protective purposes.

resin canal, resin duct One of the minute tubes found in the leaves and sometimes wood of conifers that contain resin-producing cells. **[268]**

resiniferous Bearing resin.

resinous Containing or bearing resin; resembling resin.

resting (see **perennating**).

resupinate Inverted, turned upside down. In the Orchidaceae, many of the flowers are twisted through 180 degrees, so that the position of the upper and lower petals is reversed. **[251]**

resupination Inversion.

reticular Net-like in appearance or construction.

reticulate Marked with a network pattern. **[189, 255]**

retinaculum The band connecting a pollinium to the corpusculum in the Asclepiadaceae. **[233]**

retrorse Curved or bent downwards or backwards, away from the apex.

retuse Slightly notched at the apex. **[195]**

revolute Rolled downwards at the margin, i.e. towards the abaxial surface. **[188]**

rhachis (see **rachis**).

rheophyte A plant confined to flowing water.

rhipidium A fan-shaped cyme, with axes only on one plane, branching alternately to one side and the other. **[206]**

rhizodermis (see **piliferous layer**).

rhizoid One of the hair-like roots on a prothallus. **[275]**

生殖，繁殖 通过现存个体产生新个体。

匍匐的 （见"匍匐的"）。

树脂 球果类植物的特殊细胞分泌的一类具黏性、不溶于水、起保护作用的芳香性物质。

树脂道 在球果类植物叶，有时木材中发现的、内含分泌树脂细胞的细管之一。[268]

具树脂的 生树脂的。

树脂的 含树脂的，生树脂的，类似于树脂的。

休眠的 （见"多年生的"）。

倒置的 倒转的，将上面转向下的。在兰科植物中，许多花扭转180°，致使上下花瓣的位置被反转。[251]

颠倒 反转。

网状的 表面呈网状的。

网状的 以网状式样标记的。[189, 255]

花粉腺 萝藦科植物连接花粉块到着粉腺的柄。[233]

向后的，倒向的 从顶端向下或向后弯曲的。

微凹的 顶端稍凹的。[195]

外卷的 边缘下卷，即至背面。[188]

轴 （见"轴"）。

流水植物，河生植物 局限生长于流动水质，如河水中的植物。

扇状聚伞花序 花轴仅在一个平面上，从一侧和另一侧互生分枝。[206]

根被皮 （见"根毛层"）。

假根 原叶体上的毛状根之一。[275]

rhizomatous Resembling or possessing a rhizome.

rhizome A root-like stem, lying horizontally on or situated under the ground, bearing buds or shoots and adventitious roots. **[152, 153]**

rhizome tuber (see **stem tuber**).

rhizophore In the genus *Selaginella*, a leaf-less branch, which arises from a fork in the stem and grows downwards, putting out roots when it reaches the ground. **[284]**

rhomboid More or less diamond-shaped, with four equal sides. **[191]**

rigid Stiff.

riparian Growing on the banks of rivers and streams.

rod (column) One of the elongated structures that form part of the exine of a pollen grain, and radiate from its base. **[212]**

root The lower portion of the axis of a plant, usually branched, that anchors it in the soil and enables it to absorb nutrients. **[145—153]**

root cap A hollow cone of cells covering the apical meristem of a root tip and protecting it from damage as it is pushed through the soil. **[148, 158]**

rootstock A frequently subterranean stem or rhizome. **[275]**

root sucker A shoot that arises from an adventitious bud on the root, either naturally or as a consequence of wounding; a haustorium. **[148]**

root tuber The swollen part of a root. **[145]**

rosette (see **leaf rosette**).

rostellum The beak-like extension of the stigma in some members of the Orchidaceae. The pouch-like base (bursicle) breaks, under pressure from a visiting insect, to expose the viscidia. **[251]**

rostrate With a beak.

rostrum A pointed, beak-like projection, e.g.

根状茎的 类似于或具有根状茎的。

根状茎 横生地面或位于地下的根状茎，其上着生芽或枝和不定根。[152, 153]

根茎块茎 （见"块茎"）。

根托 在卷柏属中，一种无叶分枝，从茎叉处生出，向下生长，到达地面时生出根。[284]

菱形的 或多或少呈菱形的，具四个相等的边。[191]

硬质的 坚硬的。

岸栖的 生长于河岸或溪流边的。

柱 从花粉粒外壁生出的、自基部呈放射状排列的伸长结构之一。[212]

根 植物体轴上通常具分枝的下部，固着到土壤上，吸收营养物质。[145—153]

根冠 覆盖在根尖顶端分生组织上的空心锥状细胞层，当推进土壤时保护根尖免遭伤害。[148, 158]

根状茎 一类地下茎或根状茎。[275]

根出条，吸根 从根的不定芽上生出的枝条，既可以自然生长，也可以是愈伤组织的产物；吸根。[148]

块根 根的膨大部分。[145]

莲座丛 （见"叶莲座状"）。

蕊喙 兰科某些植物柱头上的喙状延伸物。在来自来访昆虫的压力下，其囊状基部破裂露出粘盘。[251]

具喙的 具喙的。

喙 一类尖锐的喙状突起，如牻牛儿苗属植

the thickened base of the style in the genus *Geranium* [232], or the structure in the fruit of some members of the Compositae by which the pappus is attached to the cypsela. [235]

物加粗的花柱基部 [232]；菊科某些植物果实上附着冠毛的结构。[235]

rosulate Arranged in a rosette. [186]

莲座状的 莲座状排列的。[186]

rotate Gamopetalous, with a short tube and spreading lobes. [211]

辐状的 具短花筒和四周开展花冠裂片的合瓣花的。[211]

round Circular. [173]

圆形的 圆形的。[173]

rounded Having a more or less semicircular outline at the apex or base. [163, 194, 195]

圆形的 顶端或基部具有或多或少半圆形轮廓的。[163, 194, 195]

rubiginous Rust-coloured.

锈红色的 铁锈颜色的。

ruderal Growing in waste places; a plant that grows in waste places.

杂草的 荒地生长的；一类荒地生长的植物。

rudimentary Not fully developed, vestigial. [227]

不发育的 不完全发育的，残余的，退化的。[227]

rufous Reddish brown.

红棕色的 红棕色的。

rugose Wrinkled. [199]

具皱的 起皱的。[199]

rugulose Finely wrinkled.

微皱的 具细皱纹的。

ruminate Looking as though chewed.

嚼烂状的 看似被嚼烂的。

runcinate With sharply cut divisions directed backward towards the base. [191]

倒向羽裂的 具向后指向基部的尖锐裂片的。[191]

runner A creeping stem, rooting and giving rise to plantlets at the nodes. [154]

纤匍茎 一类节上生出不定根、长出小植株的匍匐茎。[154]

rupicolous (saxatile, saxicolous) Growing amongst rocks.

岩栖的 生于岩石丛的。

russet A variety of apple or potato which has a reddish brown, usually rough skin.

红褐色 苹果或马铃薯变种具红褐色、通常粗糙果皮。

sac A small pouch or bag-shaped structure.

囊 小袋或袋状结构。

saccate Pouched or bag-like. [211]

囊状的 袋状的。[211]

sagittate Arrowhead-shaped. [191, 194]

箭形的 箭形的。[191, 194]

salicetum A collection of *Salix* (willow) species and varieties for scientific or ornamental purposes.

柳圃 为科研、观赏目的而设立的柳属植物种及变种的收集园地。

saline Containing chemical salts, especially sodium chloride.

含盐的 含有化学物质盐类，特别是氯化钠。

salt gland One of the epidermal glands in the leaves of many salt marsh plants and mangroves that secrete sodium chloride. [197]

盐腺 多数盐生沼泽植物或红树植物叶片上能够分泌氯化钠的表皮腺体之一。[197]

salverform, salver-shaped (see **hypocrateriform**).

高脚蝶状的（见"高脚蝶状的"）。

samara A dry, indehiscent, winged fruit as in the genus *Fraxinus* (ash). **[179]**

samaroid Resembling a samara.

sap The juice of a plant.

saprophyte A plant that lives on decaying organic matter.

saprophytic Living on decaying organic matter.

sapwood The outer, younger layers of wood in the trunk or branch of a tree or in the stems of a shrub, usually paler in colour than the heartwood, and with living cells that are still capable of conducting sap. **[171]**

sarcotesta A fleshy seed coat.

sarcotestal Having a sarcotesta. **[159]**

sarmentose Bearing long, flexuous runners or stolons.

saxatile, saxicolous (see **rupicolous**).

scaberulous Slightly rough.

scabrid Somewhat scabrous.

scabrous With the surface rough to the touch, due to minute projections. **[200]**

scalariform Ladder-like, or having ladder-like markings.

scale, scale leaf A reduced leaf, usually membranous, and often found covering buds, bulbs and corms. **[150, 152, 153]**

scandent Climbing.

scape A leafless stalk, arising from the ground, which bears one or more single flowers, e.g. *Hyacinthoides* (bluebell), or a head of flowers, e.g. *Taraxacum* (dandelion). **[237]**

scapose Having a scape. **[165]**

scar The mark left when a part has fallen from its place of attachment. **[170, 243, 261]**

scarious Thin, dry and membranous.

schizocarp A fruit derived from a syncarpous ovary which breaks up at maturity into one-seeded portions (mericarps), as in most members of the Geraniaceae (5 mericarps)

翅果 一种干燥、不开裂、具翅的果实，如白蜡属植物的果实。

翅果状的 类似于翅果的。 [258]

汁、液 植物的汁液。

腐生植物 一类生长于腐烂有机物质上的植物。

腐生的 生长于腐烂有机物质上的。

边材 乔木主干或侧枝或灌木茎外面幼嫩的通常色泽比心材更浅、具有仍能输导汁液的活细胞的木质层。 [171]

浆果肉 肉质种皮。

浆果肉的 具有肉质种皮的。 [159]

长匍茎的 生有长匍茎的。

岩生的 。（见"岩生的"）。

微糙的 稍微粗糙的。

粗糙的 有点粗糙的。

粗糙的 具有因小突起而触摸有粗糙感表面的。 [200]

梯状的 梯状的，或有梯状纹饰的。

鳞片 一类通常膜质、覆盖芽、鳞茎和球茎的退化叶。 [150, 152, 153]

攀缘的 攀缘的。

花葶 一类从地面生出的无叶柄，着生一朵或多朵单花，如蓝铃花属植物（蓝铃花）；或一头状花序，如蒲公英属植物（蒲公英）。 [237]

具花葶的 具有花葶的。 [165]

痕 植物体某一部分自着生处脱落后留下的痕迹。 [170, 243, 261]

干膜质的 薄、干燥、膜质的。

分果 一类自合生心皮的子房发育而来、成熟时开裂为含1枚种子的分果爿的果实，像牻牛儿苗科（5枚分果爿）和伞形科（2枚分果爿）的绝大多数植物。 [257]

and Umbelliferae (2 mericarps). **[257]**

schizocarpic Bearing schizocarps.

scion A young shoot which is inserted into a rooted stock in grafting.

sciophilous Shade-loving.

sclereid Short, blunt-ended cells that, together with the fibres, compose the sclerenchyma.

sclerenchyma Strengthening tissue consisting of two kinds of cells (fibres and sclereids) with thick, often lignified walls, that supports the softer tissues of the plant. **[168, 197]**

sclerenchymatous Relating to or consisting of sclerenchyma.

sclerophyll A leaf that is tough and leathery due to well developed sclerenchyma, usually evergreen; a plant with such leaves, typical of trees and shrubs in warm, dry climates.

sclerophyllous Having tough, leathery, usually evergreen leaves.

sclerotic Hardened, stony in texture.

scorpioid cyme (see **cincinnus**).

scrambler A rather weak-stemmed climbing plant that sprawls over other plants, fences, and walls.

scrobiculate Having the surface pitted with small rounded depressions.

scurfy Covered with hake-like scales.

scutate Shaped like an oblong shield.

scutellum A more or less shield-shaped structure, situated between the endosperm and the embryo of a grass, that at germination secretes enzymes into and absorbs sugars from the endosperm. **[158]**

secondary thickening The secondary xylem and phloem produced by the vascular cambium. **[171]**

secretion Fluid produced by and released from a gland. **[207]**

sect. Section.

section A subdivision of a genus, e.g. *Primula*

具分果的 生有分果的。

接穗 在嫁接时插到根状砧木上的幼嫩枝条。

喜阴的 喜阴凉的。

石细胞 一类与纤维在一起组成机械组织的短而顶端钝的细胞。

厚壁的组织 一类含有两类厚而木质化细胞壁的细胞（纤维和石细胞），支持植物柔软组织的强壮组织。 **[168, 197]**

厚壁组织的 与机械组织相关的，或含机械组织的。

硬叶 一类因具发育良好的机械组织而坚韧、革质、通常常绿的叶；一类具有这样的叶、生活在温暖、干燥气候下的乔木和灌木类植物。

硬叶的 具坚韧、革质、通常常绿的叶。

硬化的 结构变硬的或石化的。

蝎尾状聚伞花序 （见"蝎尾状聚伞花序"）。

攀缘植物 一类具有细弱茎，在其他植物、篱笆和墙上蔓延的爬行植物。

具小网眼的 具小圆形凹陷表面的。

具鳞屑的 覆盖有碎片状鳞片的。

圆盾状的 形状像椭圆形盾片的。

盾片 禾本科植物颖果内位于胚乳和胚之间、萌发时分泌酶和从胚乳吸收糖类物质的多少呈盾状的结构。 **[158]**

次生加厚 维管形成层产生的次生木质部和次生韧皮部。 **[171]**

分泌物 腺体产生和释放的液体物质。 **[207]**

组 组。

组 属的次一等级，如报春花属皱叶报

(sect. Bullatae).

春组。

secund Directed towards one side. **[209]**

偏向的 偏向一侧的。 [209]

seed A unit of sexual reproduction developed from the fertilised ovule. **[157—160]**

种子 从受精卵发育而来的有性生殖单位。 [157—160]

seed apomixis (see **agamospermy**).

种子无融合生殖 （见"无配子种子生殖"）。

seed leaf (see **cotyledon**).

子叶 （见"子叶"）。

seedling A young plant that has grown from a seed. **[157—159]**

幼苗 从种子生长来的幼嫩植物。 [157—159]

seed plant A member of the Spermatophyta.

种子植物 种子植物门的植物。

segment One of the divisions of an organ. **[211, 249]**

裂片，部分 器官分裂后的部分之一。 [211, 249]

segregate species A species, distinguished often on the basis of minute characters, within an aggregate species.

分离种 在聚合种内，通常基于微小特征区别的种。

seismonasty The response of a plant to being touched, e.g. *Mimosa pudica* (Sensitive Plant). **[180]**

感震性 植物对触摸的响应，如含羞草。 [180]

self-fertilisation (see **autogamy**).

自花受精 （见"自花授粉"）。

self-incompatible Of a cultivar, producing viable pollen and a functional ovary, but unable to produce fruit when self-pollinated, although it may be effective in the pollination of another cultivar.

自交不亲和的 一品种虽产生可行花粉和具功能的子房，但自花传粉时不能产生果实，尽管对另一品种的传粉是有效的。

self-sterile Unable to produce viable progeny by self-fertilisation.

自花不育的 不能通过自花受精产生可行的后代。

semi-amplexicaul Clasping the stem, but only to a small extent. **[187]**

半抱茎 抱茎，但仅到很小程度。 [187]

semi-double flower A flower having petals or petaloids intermediate in number between the typical and the double forms. **[210]**

半重瓣花 一类具有大量花瓣和花瓣状中间物、处于典型花和重被花之间的花。 [210]

semi-inferior (half-epigynous, half-inferior) Of an ovary, the lower part of which is embedded in the pedicel but the upper part is free.

半下位（半上位、半下位） 子房下部埋在花柄内，而上部分离的。

seminal Relating to the seed.

种子的 与种子有关的。

semi-parasite (see **hemi-parasite**).

半寄生物 （见"半寄生物"）。

semi-parasitic (see **hemi-parasitic**).

半寄生性的 （见"半寄生性的"）。

senescent Becoming old, and thereby losing the power of cell division and growth.

衰老的 变老，从而失去细胞分裂和细胞生长的能量。

sens. (sensu) In the sense of.

在含义，在意义 在……意义上。

sens. lat. (sensu lato) In a broad sense.

广义的 在广义上。

sens. strict. (sensu stricto) In a narrow sense.

sepal A single segment of the calyx. **[232]**

sepaloid Sepal-like.

septate Divided into compartments by walls or partitions.

septicidal Splitting at maturity along or into the partitions (septa) of the capsule. **[219]**

septifragal With the valves at maturity breaking away from the partitions (septa) of the capsule. **[219]**

septum A partition, as that separating the loculi in an ovary. **[219, 265]**

ser. Series.

sericeous Silky. **[200]**

series A subdivision of a genus, e.g. *Iris* (ser. Laevigatae).

serotinal Occurring in late summer.

serotinous Of cones, remaining closed long after reaching maturity; of plants, releasing seeds only after burning.

serrate With a saw-toothed margin. **[193]**

serrulate Minutely serrate. **[193]**

sessile Not stalked. **[187, 221]**

seta A bristle.

setiferous Bearing bristles.

setose Bristly. **[200]**

sexaperturate Having six pores.

shade plant A plant adapted to living in low light intensities.

sheath A tubular covering. **[158]**

sheathing With at least the base of the leaf or stipule forming a tube that more or less encloses the stem or stalk. **[177, 178, 187]**

shoot A developed bud or young, green stem. **[152]**

shrub A woody, perennial plant, generally smaller than a tree, and with several stems arising from ground level. **[162]**

silicle, silicula The capsular fruit of the Cruciferae when less than three times as long as broad. **[264]**

较狭义的 在狭义上。

萼片 花萼的单个裂片。 [232]

萼片状的 萼片状的。

具隔膜的 被壁或隔分成隔间的。

室间开裂的 蒴果成熟时沿隔膜或从隔膜处开裂的。 [219]

室轴开裂的 蒴果成熟时从隔膜处裂开果瓣的。 [219]

隔膜 像子房内分割成室的隔。 [219, 265]

系 系列。

被绢毛的 绢状的。 [200]

系 属的次一等级，如鸢尾属光滑组。

晚夏的 夏季后期发生的。

迟季的 球果到达成熟很久以后仍维持闭合的；植物只有当燃烧后才释放种子的。

有锯齿的 具有锯齿状边缘。 [193]

具小锯齿的 小锯齿的。 [193]

无柄的 没有柄的。 [187, 221]

刚毛 短硬毛。

具刚毛的 具刚毛的。

具刚毛的 刚毛状的、具刚毛的。 [200]

具六孔的 有六个孔的。

阴生植物 一类适于生长在低光强度下的植物。

鞘 筒状遮盖物。 [158]

具鞘的 具有至少由叶基部或托叶形成、多少抱茎或柄的筒状物的。 [177, 178, 187]

苗 发育的芽或幼嫩的绿茎。 [152]

灌木 一般小于乔木、并从地面基部生出数条茎的木本多年生植物。 [162]

短角果 十字花科植物的长短于3倍宽的蒴果状果实。 [264]

siliqua, silique The capsular fruit of the Cruciferae when at least three times as long as broad. **[264]**

长角果 十字花科植物的长至少为3倍宽的蒴果状果实。 [264]

siliquiform In the form of a siliqua.

长角状的 长角果状的。

silks The long, silk-like styles in the female inflorescence of *Zea mays* (Maize, Sweet Corn). **[244]**

穗丝 玉米雌花序上的细长而丝状的花柱。 [244]

simple Of one piece. **[190, 191, 202, 204, 205]**

单的 单片的。 [190, 191, 202, 204, 205]

single flower A flower having the typical number of petals or petaloids. **[210]**

单被花 一类具有典型数目的花瓣或花瓣状物的花。 [210]

sinker A shoot that grows downwards from a bulb or a corm and produces a bulb or a corm at its apex **[154]**; in parasitic plants, an outgrowth of the haustorium that extends into the tissues of the host plant. **[147]**

茎条，次生吸根 从鳞茎或球茎上向下生长的、并在其末端再次生出鳞茎或球茎的枝条 [154]；在寄生植物中，伸向寄主植物组织的吸根生长物。 [147]

sinuate Having the blade of the leaf flat but with the margin winding strongly inward and outward. **[193]**

具深波状的 具有扁平叶片但边缘向内向外强烈弯曲的。 [193]

sinus The recess between two teeth or lobes, e.g. on a leaf margin. **[193]**

凹缺 如叶缘上两齿或裂片之间的弯缺。 [193]

siphonostele A type of stele in which the xylem and phloem surround a central core of pith. **[280]** (see **amphiphloic** and **ectophloic**)

管状中柱 一类木质部和韧皮部围绕着中心髓的中柱 [280]（见"双韧的"和"外韧的"）。

s.1. (sensu lato) In a broad sense.

广义的 在广义上。

sleep movement (see **nyctinasty**)

睡眠运动 （见"感夜性"）。

soboliferous Forming clumps.

具根出条的 形成丛的。

sodium chloride (NaCl) Common salt.

氯化钠 普通食盐。

softwood Wood obtained from conifers.

针叶材 从松柏类植物获得的木材。

solenostele (see **amphiphloic siphonostele**)

管状中柱 （见"双韧管状中柱"）。

solitary Borne singly.

单生的 单生的。

solute A substance that is dissolved in water or another liquid.

溶质 一类溶解到水或另一液体中的物质。

soriferous Bearing sori.

具狍子囊群的 生有孢子囊群的。

sorosis (plural **soroses**) A fleshy, multiple fruit, e.g. *Ananas* (pineapple) or *Morus* (mulberry). **[262]**

椹果（复数为soroses） 一类肉质复合果，如菠萝属（菠萝）和桑属（桑椹）。 [262]

sorus A group of sporangia. **[275]**

囊群 孢子囊群。 [275]

sp. Species (singular).

种 种（单数）。

spadix (plural **spadices**) A spike with a fleshy axis, as in Araceae. **[207]**

肉穗花序（复数为spadices） 具有肉质花序轴的穗状花序，如天南星科植物。 [207]

spathe A large bract subtending and often enclosing a flower or an inflorescence. **[155, 207]**

佛焰苞 通常包着一朵花或一枚花序的大苞片。[155, 207]

spathulate, spatulate Spatula-shaped. **[191, 233]**

匙形的 匙形的。[191, 233]

species (plura **species**) A group of closely related, mutually fertile individuals, showing constant differences from allied groups, the basic unit of classification. The names of species are written nowadays with a small initial letter, but formerly the names of some species, e.g. those derived from personal names, were written with a capital initial letter.

种（复数为species） 一群关系紧密、彼此间可育的个体，显示出与近缘种稳定的区别，分类的基本单位。种名现在第一字母用小写，但以前一些种名，如来源于人名的，第一字母用大写。

specific epithet The second element, usually an adjective, in the binomial name of a plant, which follows the generic name and serves to distinguish one species from others in the same genus.

种加词 植物双名名称的第二成分，通常为形容词，跟在属名之后，用于区别一个种与该属其他种。

Spermatophyta, spermatophytes Seed-bearing plants, comprising the Angiospermae and the Gymnospermae.

种子植物门、种子植物 产生种子的植物，包括被子植物和裸子植物。

spermatozoid (see **antherozoid**).

游动精子 （见"游动精子"）。

sperm cell (see **male nucleus**).

精细胞 （见"雄核"）。

spicate Spike-like. **[204]**

穗状花序上 穗状花序状的。[204]

spicule A small or secondary spike.

小穗 小的或次生的穗状花序。

spike An indeterminate inflorescence with sessile flowers. **[204]**

穗状花序 一类具有无柄花的无限花序 [204]

spikelet A unit of the inflorescence in grasses, consisting of one or more flowers subtended by a common pair of glumes. **[243]**

小穗 禾本科植物花序的组成单位，由包在一对常见颖片内的一或多朵花组成。[243]

spine A sharp woody or hardened outgrowth from a leaf, sometimes representing the entire leaf, or from a fruit. **[180, 182, 240, 256]**

刺 从叶上（有时代表整个叶），或果实上生出的一类尖锐、木质或变硬的生长物。[180, 182, 240, 256]

spinescent Ending in a spine or sharp point.

具刺的 终止于刺或尖点。

spinose Spiny. **[193]**

具刺的 有刺的。[193]

spinule A small spines.

小刺 小的刺。

spinulose Bearing small spines.

具微刺的 生有小刺的。

spiralled cyme (see **cincinnus**).

螺旋状聚伞花序 （见"蝎尾状聚伞花序"）。

spirally arranged With the leaves arranged alternately along a stem so that a line joining

螺旋状排列的 叶沿茎互生排列以致连接着生点的线形成螺旋。[175]

their points of attachment would form a spiral. **[175]**

spongy mesophyll The tissue with a conspicuously open texture lying beneath the palisade layer in the leaves of mesophytic dicotyledons, and consisting of cells with large intercellular spaces. **[196]**

sporangiophore The stalk of a sporangium, as in the genus *Equisetum* (horsetail). **[283]**

sporangium A spore-case, the structure in which the spores are produced. **[275]**

spore The reproductive unit in ferns, produced in the sporangium and developing into a prothallus. **[275]**

sporeling A very young fern plant that has developed from the prothallus. **[275]**

sporiferous Bearing spores.

sporocarp The body enclosing the sporangia, especially the hard, rounded structure near the base of the leaves in some aquatic ferns, e.g. *Pilularia*. **[281]**

sporophyll A specialised leaf that bears sporangia. **[282]**

sporophyte The mature stage in the life-cycle of ferns when the cells are diploid or, more rarely, polyploid. **[275]**

sport (see **mutant**).

spp. Species (plural).

spring wood (early wood) Pale-coloured wood, with large xylem cells, that is produced early in the growing season. **[171]**

sprout A bud or newly grown shoot.

spur A slender projection from a plant part, especially the tubular extension, often nectariferous, of the calyx or corolla that occurs in *Aconitum*, *Aquilegia*, *Linaria* *Viola* etc. **[226]**; a short, leafy branch of a tree, often with flowers and fruits in clusters at closely-spaced nodes. **[261]**

squamiform Scale-like, as the adult leaves of *Juniperus*. **[269]**

叶内 中生双子叶植物叶的位于栅栏组织下面的具明显开放结构、包括发达细胞间隙细胞的组织。[196]

孢囊柄 像木贼属植物（木贼）的孢子囊的柄。[283]

孢子囊 产生孢子的结构。[275]

孢子 蕨类植物在孢子囊内产生的、发育成原叶体的生殖单位。[275]

萌芽孢子，孢苗 从原叶体发育来的非常幼小的蕨类植物。[275]

产孢子的 着生孢子的。

孢子果 尤其是某些水生蕨类，如线叶萍属植物叶近基部的坚硬、圆形结构，里面包含孢子囊。[281]

孢子叶 具有孢子囊的特化叶。[282]

孢子体 蕨类植物生活史中的成熟阶段，细胞为二倍体，或多稀多倍体。[275]

突变 （见"突变体"）。

种 种（复数）。

春材（早材） 在生长季早期产生的、具有大的木质部细胞的浅色的木材。[171]

抽芽，籽苗 芽或新生枝。

距，短枝 一类从植物部分的纤细突起物，特别是乌头属、耧斗菜属、柳穿鱼属和堇菜属等植物的花萼或花冠的管状延伸，通常具有蜜腺 [226]；树木的具叶短枝，在密集结上簇生着花和果实。[261]

鳞片状的 像刺柏属植物的成年叶一样，鳞片状的。[269]

squamose Covered with large, coarse scales. **[199]**

squamulose Covered with small scales.

square Having four sides. **[173]**

squarrose With a rough surface due to projecting hairs or scales **[199]**; in trees and shrubs, with branches projecting more or less at right-angles to the main stem.

squarrulose Minutely squarrose.

ssp. (plural **sspp.**) Subspecies.

s.str. (sensu stricto) In a narrow sense.

stamen One of the male sex organs, usually consisting of anther, connective, and filament. **[207, 214—218]**

staminal Relating to stamens. **[231]**

staminate Having only male organs.

staminodal (staminodial) Relating to staminodes.

staminode A sterile stamen. **[217, 236, 251]**

staminodial (see **staminodal**).

standard The large upper petal (vexillum) in the flowers of plants of subfamily Papilionoideae in the Leguminosae **[230]**; one of the three more or less erect inner perianth segments in flowers of the genus *Iris* **[238]**; a tree, especially a fruit tree, trained so that it has an upright stem free of branches.

starch grain A rounded or irregular mass of starch within a chloroplast or other plastid.

stele (vascular cylinder) The central core of the stem and root in a vascular plant, consisting of vascular tissue and often associated tissue such as pith, pericycle etc. **[148, 168]**

stellate Star-shaped. **[201]**

stellular, stellulate Shaped like a small star.

stem The main supporting axis of a plant. **[151, 153, 168]**

stem leaf A leaf borne on the stem as opposed to being at its base.

具鳞片的 被大而糙的鳞片。[199]

具小鳞片的 被小鳞片的。

正方形的 具有四边的。[173]

具糙鳞的，糠秕状的 具突出毛或鳞片的粗糙表面的 [199]；具与主杆多少呈直角生出分枝的乔木和灌木。

小糠秕状的 小糙鳞的。

亚种 亚种（复数为sspp.）。

狭义的 在狭义上。

雄蕊 雄性器官之一，通常包括花药、药隔和花丝。[207, 214—218]

雄蕊的 与雄蕊有关的。[231]

雄性的 仅具有雄性器官的。

退化雄蕊的 与退化雄蕊有关的。

退化雄蕊的 不育雄蕊。[217, 236, 251]

退化雄蕊的 与不育雄蕊相关的。

旗瓣 豆科蝶形花亚科植物花的大的上面花瓣 [230]；鸢尾属植物花的三枚多少直立的内花被裂片之一 [238]；一类树木，特别是果树，经过驯化，以致它具有一个没有分枝的直的茎干。

淀粉粒 叶绿体或其他质体内的圆形或不规则的淀粉团。

中柱（维管柱） 维管植物茎和根的中心柱，包括维管组织和通常的辅助组织，如髓、中柱鞘等。[148, 168]

星状的 星状的。[201]

小星状体的 形状像小星星的。

茎 植物体的主要支撑轴。[151, 153, 168]

茎生叶 与基生叶相对应的生长在茎上的叶

stem tuber (rhizome tuber) The swollen end of an underground stem, e.g. a potato tuber. **[152]**

sterile Unable to reproduce sexually. **[207]**

stigma The apex of the style, usually enlarged, on which the pollen grains alight and germinate. **[213, 219—221]**

stigma flap In the genus *Iris*, a small flap on the underside of the style branch. **[238]**

stigmatic Relating to the stigma. **[225]**

stilt root (see **prop root**).

stimulus A substance or action that causes a response in the plant. **[173]**

stinging hairs Stiff, tubular hairs, filled with irritant substances, as in *Urtica dioica* (Stinging Nettle). When the tip of the hair is broken off, e.g. by an animal, pressure on the saccate base of the hair forces the contents out into the animal's skin. **[202]**

stipe A stalk, especially the caudicle in an orchid flower, or the petiole of a fern or palm frond. **[249, 252, 265, 275]**

stipel The stipule of a leaflet.

stipitate Having a stipe, or borne upon one.

stipular Relating to a stipule. **[179, 180]**

stipulate Having stipules. **[187]**

stipule A leafy outgrowth, often one of a pair, arising at the base of the petiole. **[169, 178—180, 182, 187]**

stock The rooted stem into which the scion is inserted when grafted.

stolon A lateral stem growing horizontally at ground level, rooting at the nodes and producing new plants from its buds, as in the genus *Fragaria* (strawberry).

stoloniferous Bearing stolons. **[162]**

stoma (plural stomata) One of the small pores, found most often in the epidermis on the lower surface of leaves but also on young stems, which allow gases to pass in and out of the plant. **[196, 268]**

块茎（根状块茎） 地下茎的膨大末端，如马铃薯的块茎。[152]

不育的 不能有性繁殖的。[207]

柱头 花柱通常膨大的、花粉粒降落和萌发的顶端。[213, 219—221]

柱头瓣 鸢尾属植物花柱分枝下面的小瓣。[238]

柱头的 与柱头有关的。[225]

支柱根 （见"支柱根"）。

刺激物 一类引起植物响应的物质或活动。[173]

螫毛 一类像异株荨麻植株上的充满刺激物的硬管状毛。当毛的顶端被动物折断时，挤压毛的囊状基部，迫使内容物进入动物的皮肤。[202]

柄 梗，尤其像兰花的花粉块柄，或蕨类和棕榈类叶的柄。[249, 252, 265, 275]

小托叶 小叶的托叶。

具有柄的 具柄或带有柄的。

托叶的 与托叶相关的。[179, 180]

具托叶的 具托叶的。[187]

托叶 叶柄基部生出的、通常一对之一的叶状生长物。[169, 178—180, 182, 187]

砧木 嫁接时接穗插入的具根的茎。

匍匐茎 一类在地面上水平生长的侧生茎，节处生根，从芽上生出新植株，像草莓属植物（草莓）。

具匍匐茎的 生于匍匐茎的。[162]

气孔（复数为stomata） 最通常情况下在叶下表面表皮及嫩茎上发现的、有利于气体出入植物体的小孔之一。[196, 268]

stomatal Relating to stomata.

stomium The opening in the annulus of the sporangium through which the spores are released. **[275]**

stone (see **pyrene**).

stone cell A strongly lignified type of sclereid found in the flesh of fruits in the genus *Pyrus* (pear).

storage organ The swollen part of a plant where food is stored in the form of starch or sugar, e.g. a tuber, bulb, or corm. **[150—153]**

stramineous Straw-coloured.

striate Having fine, longitudinal lines, grooves, or ridges.

strigose Bearing stiff hairs or bristles. **[200]**

strobile, strobilus (cone) The reproductive structure in gymnosperms **[283]**. In conifers, this consists of an ovoid, cylindrical, or spherical cluster of sporophylls (cone scales) arranged round a central axis [267]. The term is sometimes applied to e.g. the papery, cone-shaped fruits of *Humulus lupulus* (Hop). **[262]**

strobiloid Resembling a strobilus.

stroma The colourless material inside a chloroplast in which the chlorophyll necessary for photosynthesis is embedded.

stylar Relating to the style. **[238]**

style The often elongated apical part of a carpel or gynoecium that bears the stigma at its tip. **[158, 207, 213, 219—221]**

stylopodic Arising from an enlarged base. **[220]**

stylopodium The enlargement at the base of the styles in some members of the Umbelliferae. **[220]**

subapical Almost at the apex.

subbasal Almost at the base.

subclass A subdivision of a class. Names of subclasses end in '-idae'.

subcompound More or less compound.

subcordate More or less heart-shaped. **[194]**

气孔的 与气孔有关的。

裂口 孢子囊环带上释放孢子的裂口。[275]

果核 （见"果核"）。

石细胞 一类在梨属植物（梨）肉质果实中发现的强烈木质化的石细胞。

贮藏器官 植物以淀粉或糖类形式贮藏营养物质的膨大部分，如块茎、鳞茎或球茎。[150—153]

麦秆色的 淡黄色的。

具条纹的 具有细纵线、沟或脊。

具糙状毛的 具硬毛或刚毛的。[200]

孢子叶球（球果） 裸子植物的繁殖结构[283]。在球果类，该结构包括由中轴周围排列的孢子叶（果鳞）构成的卵形、柱状或球形簇 [267]。这一术语有时用于啤酒花（忽布）的纸质、球果状果实。[262]

球果状的 类似于球果的。

基质 叶绿体内的无色物质，光合作用进行需要叶绿素。

花柱的 与花柱有关的。[238]

花柱 顶端着生柱头的心皮或雌蕊群的通常伸长的顶端部分。[158, 207, 213, 219—221]

花柱基的 从膨大的花柱基部生出的。[220]

花柱基 伞形科某些植物花柱基部膨大物。[220]

近尖端的 几乎在顶端的。

近基部的 几乎在基部的。

亚纲 纲的次一等级，亚纲的名称以-idae结尾。

近复合的 多少复合的。

近心形的 多少心状的。[194]

subcuneate More or less wedge-shaped.

subdivision The rank immediately below division. Names of subdivisions end in '-phytina'.

subequal Almost equal.

suberin A mixture of fatty substances present in the cell walls of cork that renders them waterproof and resistant to decay.

suberised Converted into cork.

suberose Corky.

subfamily A subdivision of a family. Names of subfamilies end in '-oideae'.

subform, subforma The lowest taxonomic rank, a subdivision of a form or forma.

subg. Subgenus.

subgenus (plural subgenera) A subdivision of a genus, e.g. *Prunus* (subg. Cerasus).

subglabrous Almost without hairs.

subglobose (depressed-globose, oblate) Almost globular, but flattened at the ends of the axis.

subkingdom A subdivision of kingdom, a taxonomic rank.

suborbicular Almost circular.

suborder A subdivision of an order. Names of suborders end in '-ineae'.

subpetiolar, subpetiolate Under the petiole, e.g. in the genus *Platanus* (plane), the petiole is expanded at the base to form a hood over the axillary bud. **[177]**

subsessile Almost devoid of a stalk.

subshrub (suffrutex, nudershrub) A low shrub, sometimes with partially herbaceous stems. **[162]**

subsp. (plural subspp.) Subspecies.

subspecies (plural subspecies) A subdivision of a species, often used for a geographically or ecologically distinct group of plants.

subtend To stand below and close to, to extend under. **[184]**

subtribe A subdivision of a tribe. Names of

近楔形的 多少楔状的。

亚门 门的次一等级，亚门的名称以-phytina 结尾。

近相等的 几乎相等的。

木栓质 一类存在于木栓细胞壁内、使其防水抵抗腐烂的脂类物质混合物。

栓化的 转化成木栓。

栓质的 木栓的。

亚科 科的次一等级，亚科的名称以-oideae 结尾。

亚变型 变型的次一等级，最低分类等级。

亚属 亚属。

亚属（复数为subgenera） 属的次一等级，如李属梅亚属（*Prunus* subg. *Cerasus*）

近无毛的 几乎没有毛的。

近球形的（扁球形的，扁圆形的） 几乎球形的，但轴端扁平。

亚界 界的次一等级，一种分类等级。

近圆形的 几乎圆形的。

亚目 目的次一等级，亚目的名称以-ineae 结尾。

叶柄下的 如悬铃木属植物（悬铃木），叶柄基部扩大形成帽状，罩在腋芽上面。[177]

近无柄 叶柄几乎缺少。

亚灌木 一类低矮灌木，有时具有部分草质茎。[162]

亚种 亚种（复数为subspp.）。

亚种（复数为subspecies） 种的次一等级，通常用于一类具有地理或生态区别的植物群。

包着 位于下方并贴近，向下延伸。[184]

亚族，族的次一等级，亚族的名称以-inae

subtribes end in '-inae'.

subulate Awl-shaped, tapering from the base to the apex. **[191]**

subvalvate Almost valvate.

subvar. Subvariety.

subvariety A subdivision of a variety.

subvars. Subvarieties.

succulent Fleshy and juicy.

sucker A shoot of subterranean origin. **[154]** (see also **root sucker**)

sucker disc The expanded tip of some tendrils that enables them to adhere to fences, walls etc., e.g. *Parthenocissus* (Vitaceae). **[179]**

suffrutescent Woody only at the base of the stem.

suffrutex (see **subshrub**).

suffruticose Woody in the lower part of the stem.

sulcate Grooved or furrowed.

sulcus A groove or furrow.

summer wood (see **autumn wood**).

superior Above, as when the ovary is situated above the other floral parts on the receptacle. **[218]**

superposed Placed above or on another structure.

supervolute In ptyxis, having one margin rolled within the other. **[188]**

supra-axillary Growing above an axil.

suprafoliar Borne above the leaves. **[247]**

surcurrent Running up, as when the base of the leaf is prolonged up the stem as a wing. **[278]**

surmounted Capped or crowned. **[219]**

suspensor In spermatophytes and certain pteridophytes, the group of cells within the embryo sac which pushes the embryo down into the developing endosperm. **[284]**

suture A seam or line of joining. **[219, 256]**

switch plant A plant of dry places with long, thin stems, which at first bear a few leaves

结尾。

钻形的 锥形，从基部到顶端逐渐变尖。[191]

近镊合状的 近镊合状的。

亚变种 亚变种。

亚变种 变种的次一等级。

亚变种 亚变种（复数）。

肉质的，多汁的 肉质的，多汁的。

根出条 地表下面起源的枝条 [154]（也见"根出条"）。

吸盘 能使其粘贴在栅栏、墙上的某些卷须的膨大顶端，如爬山虎属（葡萄科）植物。[179]

半灌木状的 仅茎基部木质的。

亚灌木 （见"亚灌木"）。

半灌木状的 茎下部木质的。

具深沟的，具槽的 具沟的或槽的。

沟槽 沟或槽的。

晚 （见"秋材"）.

上位的，上面的 像子房位于花托上花的其他部分之上一样。[218]

叠生的 置于另一结构上或之上。

重席卷的 在叶片卷迭式中，一侧边缘卷到另一侧边缘内的。[188]

腋上生的 叶腋以上生长的。

生于叶面的 生于叶上的。[247]

翅状伸张的 像叶基部以翅的形式延伸到茎一样。[278]

冠以 加帽的，具冠的。[219]

胚柄 种子植物和某些蕨类植物的胚囊内将胚推进发育胚乳的细胞群。[284]

缝 接缝或接合线。[219, 256]

转换植物 一类具有细长茎的干旱生境的植物，起初生出少量叶，但接下来变绿，

but subsequently, being green, take over the process of photosynthesis.

syconium A multiple hollow fruit, as in the genus *Ficus* (fig). **[263]**

syllepsis Growth of a bud into a lateral shoot without any period of dormancy.

sylleptic Growing from a bud into a lateral shoot without any delay.

symbiont Either one of a pair of organisms involved in symbiosis.

symbiosis The arrangement whereby two different organisms (symbionts) co-exist, not necessarily to their mutual advantage, though often used in this respect. (see also **commensalism** and **mutualism**)

sympatric Of plant species or populations, having a common or an overlapping distribution.

sympetalous (see **gamopetalous**).

sympodial With the main stem or axis ceasing to elongate but growth being continued by the lateral branches.

syn. Synonym.

synandrium An androecium coherent by the anthers, as in some members of the Araceae.

synandrous With coherent anthers.

synangium A compound structure formed by the fusion of two or more sporangia in certain tropical ferns, e.g. *Platycerium*, and fern allies, e.g. *Psilotum*, or by the fusion of groups of pollen sacs in certain cycads. **[281]**

syncarp, syncarpium A fleshy, multiple fruit with united carpels. **[262]**

syncarpous Having united carpels.

synergid One of the two haploid nuclei that, together with the ovum, lie at the micropylar end of the embryo sac. **[213]**

synflorescence A compound inflorescence, composed of a terminal inflorescence (florescence) and one or more lateral inflorescences (co-florescences).

进行光合作用。

隐头花序（隐头果） 一类复合的空心果实，像榕属植物（无花果）的果实。[263]

同期生长 不经过任何休眠期，从芽到侧枝的生长。

同期生长的 没有任何拖延，从芽到侧枝的生长。

共生生物，共生体 参与共生的一对有机体之一。

共生，共栖 通过两种不同生物共存的安排，尽管经常用在这方面，但对它们彼此进步不是必要的。（也见"偏利共生"和"互惠共生"）

同域的，分布区重叠的 拥有共同或重叠分布区的植物种或居群的。

合瓣的 （见"合瓣的"）。

合轴的 主茎或轴停止生长，依靠侧枝继续生长。

异名 异名。

聚药 一类花药黏合的雄蕊群，如天南星科的某些植物。

聚药的 具有黏合花药的。

聚合囊 某些热带蕨类如鹿角蕨属和拟蕨类如松叶蕨属的两个或多个孢子囊愈合，或某些苏铁类的花粉囊群愈合形成的复合结构。[281]

合心皮果 具有合生心皮的肉质复果。[262]

合心皮的 具有合生心皮的。

助细胞 和卵细胞一起，位于胚囊珠孔端的两个单倍核之一。[213]

复合花序 由一枚顶生花序和一或多枚侧生花序构成的复合花序。

syngamy Sexual reproduction involving the fusion of a male and a female gamete.

融合生殖，配子配合 涉及雄配子和雌配子融合的有性生殖。

syngenesious With anthers united into a tube, but filaments free, as in flowers of the Compositae. **[215]**

聚药的 像菊科植物那样，花的雄蕊具有结合成筒状的花药，但花丝离生。 [215]

synonym Another name for the same taxon, either an alternative name that is valid under a different classification, or a name now invalid according to the I.C.B.N. that has been superceded by a later name.

异名 同一类群的另一名称，既可以是不同分类系统下的有效的互用名称，也可以是现在根据《国际植物命名法规》已经被后来名称替代的无效名称。

synsepal (see **synsepalum**).

合萼 （见"合萼"）。

synsepalous (see **gamosepalous**).

合萼的 （见"合萼的"）。

synsepalum (synsepal) A structure formed by the joining of two or more sepals, especially in certain genera of the Orchidaceae, e.g. in *Masdevallia* all three sepals are connate to some extent, and in *Paphiopedilum* and most species of *Cypripedium* the two lateral sepals are united. **[251]**

合萼 由2枚或更多枚萼片合生而成的结构，特别是兰科一些属植物，如三尖兰属的3枚萼片都一定程度的合生，兜兰属和杓兰属的绝大多数植物的2枚侧生萼片合生。 [251]

syntepalous Having the tepals united.

合被的 具合生花被片的。

systematic botany (see **taxonomy**).

系统植物学 （见"分类学"）。

t., tab. (tabula) A plate, a full-page illustration.

图版 图版，全页插图。

tannin An acidic substance, soluble in water, with a bitter taste and astringent properties, that is present in a number of plants, especially in the bark of species of *Quercus* (oak).

丹宁 一种溶于水、有苦味和收敛性的酸性物质，存位于大量植物中，特别是栎属植物（橡树）的树皮内。

tanniniferous Bearing tannin.

含丹宁的 生有丹宁的。

tapetam The layer of cells that forms a nutrient tissue round the pollen mother-cells in the anthers of flowering plants or the spore mother-cells in pteridophytes. **[212]**

绒毡层 有花植物花药内花粉母细胞周围或蕨类植物的孢子母细胞周围的形成营养组织的细胞层。 [212]

taproot A strongly developed main root which grows downwards bearing lateral roots much smaller than itself. **[145, 151, 153, 169]**

直根 一类强烈发育的、向下生长的主根，其上生长着远比它小的侧根。 [145, 151, 153, 169]

tassel The panicle comprising the male inflorescence in *Zea mays* (Maize, Sweet Core). **[244]**

雄穗 构成玉米（玉米，甜玉米）雄花序的圆锥花序。 [244]

taxon (plural taxa) A unit of classification of any rank, e.g. *Bellis perennis* (species);

分类单位，分类群（复数为taxa） 任一等级的分类单元，如雏菊（种）、雏菊

Bellis (genus); Compositae (family).

taxonomy (systematic botany) The aspect of botany that deals with the identification, classification and nomenclature of plants.

tectum The structure forming a roof over the rods in the exine of a pollen grain. **[212]**

tendril A twining, thread-like structure produced from a stem or leaf that enables a plant to hold its position securely. **[177, 179]**

tension wood A kind of reaction wood found on the upper sides of the branches and inclined trunks of hardwood trees, and characterised by a greatly thickened inner layer of the cell walls, which may separate from the rest of the cell wall or may enlarge to fill the lumen.

tenuinucellate Having a nucellus composed of little more than an epidermis and an embryo sac.

tepal One of the petals or sepals of a flower in which all the perianth segments closely resemble each other. **[168]**

terete Like a slender, tapering cylinder, and more or less circular in any cross-section. **[220]**

terminal At the apex or end. **[150, 170, 184, 270]**

ternate In threes.

terrestrial Growing on the ground.

tessellate, tessellated Chequered, having markings or colours arranged in squares. **[189]**

testa The seed coat, a hard covering that has developed from the integument of an ovule after fertilisation. **[157]**

testiculate Shaped like the tubers of certain terrestrial orchids, e.g. *Ophrys*. **[145]**

tetrad A group of four pollen grains. **[252, 254]**

tetradynamous Having four long stamens and two short ones, as in the Cruciferae. **[215]**

tetrahedral Four-sided, as a pyramid.

tetramerous Having the parts of the flower in

属、菊科。

分类学（系统植物学） 处理植物鉴定、分类和命名的植物学分支学科。

覆盖层 形成花粉粒外壁柱顶部覆盖层的结构。[212]

卷须 一种从茎或叶演变而来的、能使植物牢固地固着其位置的缠绕性丝状结构。[177, 179]

应拉木，伸张木 一类发现于树枝和硬木树倾斜树干上侧面的畸形木材，以极度加厚的细胞壁内层为特点，该层可与细胞壁其余部分分开或扩大以填充细胞腔。

薄珠心的 具由像表皮一样的被层和一个胚囊构成的珠心的。

花被片 所有花被裂片彼此相似的花的花瓣或萼片之一。[168]

圆筒形的，圆柱状的 细长、锥形的筒，近圆形的截面。[220]

顶生的 在顶端或末端的。[150, 170, 184, 270]

三出的 三个一组的。

陆生的 生长在陆地上的。

具网格斑纹的 格子的，具以方格排列的斑点或颜色的。[189]

种皮 种子的被层，由胚珠受精后的珠被发育而成的坚硬的覆盖物。[157]

睾丸状的 形似某些陆生兰花，如欧洲对叶兰属植物块茎的。[145]

四分体 一组具四枚花粉粒。[252, 254]

四强雄蕊的 具四枚长雄蕊和两枚短雄蕊的，如十字花科植物。[215]

四面体的 四面的，如金字塔的。

四基数的 花的各部分以四为基数的。

fours.

tetrandrous Having four stamens.

tetraploid (4n) Having four sets of chromosomes in each cell.

tetrarch root A root with four protoxylem strands in the stele. **[149]**

thalamus (see **receptacle**).

thallus A plant body not differentiated into leaves and stem. **[275]**

theca (anther cell) One of the usually two lobes of an anther in which pollen is produced. At first, each anther cell is divided into two portions (pollen sacs), making four in all, but before anthesis the tissue separating each pair disintegrates, and the anther then becomes two-celled. In e.g. the family Malvaceae, the anther has only one lobe. **[216, 251]**

thermonasty The response of a plant to a change in temperature. **[174]**

thermotropic Turning towards the heat source (positively thermotropic) or away from it (negatively thermotropic).

thermotropism The movement of a plant in response to heat from a particular direction.

therophyte An annual, a plant which survives the unfavourable season of the year in the form of seeds.

thigmotropic (haptotropic) Of a plant, turning towards the object with which it has come into contact (positively thigmotropic), as many climbing stems and tendrils, or away from it (negatively thigmotropic).

thigmotropism The movement of a plant in response to physical contact with another object.

thorn A short, pointed branch. **[180]**

throat The region of a corolla or calyx with united segments where the lower, tubular portion expands into the upper, spreading

四雄蕊的 具有四枚雄蕊的。

四倍体的 每一细胞中有四套染色体的。

四原型根 中柱中具四枚原生木质部脊的根。[149]

花托 （见"花托"）。

原植体 不分化为叶和茎的植物体。[275]

囊室 组成产生花粉的花药的两部分（药室）之一。首先每一药室分为两部分（花粉囊），共形成四个花粉囊，但到开花前分隔每对花粉囊的组织解体，然后花药变为两室。如在锦葵科植物中，花药只有一个花粉囊。[216, 251]

感热性 植物对温度变化的响应。[174]

向温的 转向热源（正向热的），或离开热源（负向热的）。

向温性 植物对来自一特定方向的热源做出响应的运动。

一年生植物 以种子度过不利季节的一年生植物。

向触的 指植物体转向与它接触的物体（正向触的），如缘缘的茎和卷须，或离开接触的对象（负向触的）。

向触性 植物对与另一物体接触做出响应的运动。

刺 一类短而锐尖的枝。[180]

喉部 花冠或花萼具有合生裂片的区域，在此处下部管状部分扩展成上部开展部分。[208]

part. **[208]**

thrum-eyed One of the two forms of a dimorphic flower, e.g. *Primula vulgaris* (Primrose), where the style is short and the stamens are above the stigma. **[229]** (see also **pin-eyed**)

花药上位的，花药可见的（线式型的） 如报春花的两型花的两种形式之一，其花柱很短，雄蕊花药在柱头之上。 [229]（又见"针式型的"）

thyrse A mixed inflorescence in which the main axis is indeterminate and the secondary and ultimate axes are determinate or cymose. **[206]**

聚伞圆锥花序 一种复合花序，其主轴是无限的，次级轴或末级轴是有限的或聚伞状的。 [206]

thyrsoid Resembling a thyrse.

聚伞圆锥花序状的 类似于聚伞圆锥花序的

tiller A lateral shoot arising at ground level from the stems of grasses. **[241]**

分蘖 禾本科植物茎在地面处生出的侧枝。 [241]

tissue An aggregation of cells with a similar form and function that form the material of which a particular part of a plant is composed.

组织 一种形成组成植物体特定部分的原料、具相似形态和功能的细胞集和体。

tomentose Densely covered in soft hairs. **[200]**

被绒毛的 密被柔软毛的。 [200]

tomentum The dense covering of hairs on a plant or a particular organ.

绒毛层 在植物体或特定器官上密集的毛覆盖物。

tooth One of the small, pointed projections that form the margin of many leaves, or sometimes the apex of a mature capsule. **[256]**

齿 一类形成许多叶边缘或有时成熟蒴果顶端的小而尖锐的突起。 [256]

topocline A type of cline relating to variations in a particular taxon throughout its geographical range.

地理渐变群，地区渐变群 一种与特定类群在其地理区域内变异相关的渐变群的形式。

topodeme A deme occupying a particular geographical area.

区域同类群 占据一特定地理区域的同类群。

topogamodeme A gamodeme occupying a precise locality.

区域交配同类群 占据一准确地域的交配同类群。

torose Cylindrical, with swellings or contractions at intervals.

念珠状的 柱状但间隔处具肿胀或收缩的。

torulose Diminutive of torose. **[265]**

小念珠状的 小念珠状的。 [265]

torus (see **receptacle**).

花托 （见"花托"）。

trabecula A rod-like structure, e.g. one of the strands of sterile tissue dividing the cavity in the sporangia of *Isoetes* (quillwort). **[281, 284]**

横隔片 一种棒状结构，如水韭属植物（水韭）孢子囊中分开腔穴的不育组织束之一。 [281, 284]

trace elements (see **micronutrients**).

痕量元素 （见"微量元素"）。

tracheid An elongated structure derived from a single cell that develops lignified walls and is able to conduct water and solutes in gymnosperms and other woody plants.

Tracheophyta, tracheophytes (see **vascular plants**).

translator The clip (corpusculum) and bands (retinacula) which connect the pair of pollinia in the Asclepiadaceae. **[233]**

translocation Transport of dissolved substances within the xylem and phloem of a plant.

translucent Allowing light to pass through, but not transparent.

transpiration The loss of water vapour into the atmosphere through the stomata.

transverse Crosswise.

trapeziform, trapezoid Shaped like a trapezium, i.e. with only two of its four sides parallel.

traumatonasty The movement of a plant in response to wounding.

tree A woody, perennial plant, usually tall, with a single bole or trunk that bears a crown of branches. **[162]**

tree canopy The cover formed by the crowns of the tallest trees in a wood or forest.

tree ring (see **annual ring**).

triadelphous With three groups of stamens, as the flowers in some species of *Hypericum*.

triandrous Having three stamens, as in the Iridaceae.

triangular Having three angles and sides. **[173, 258]**

triaperturate Having three pores.

triarch root A root with three protoxylem strands in the stele. **[149]**

tribe A subdivision, usually of a large subfamily, but sometimes a family may be divided directly into tribes. Names of tribes end in '-eae'.

管胞 在裸子植物和其他木本植物中，由发育成木质细胞壁的单个细胞演化而来的、能运输水分和溶质的伸长结构。

维管植物门，维管植物 （见"维管植物"）。

载粉器 萝藦科植物连接一对花粉块的着粉腺。[233]

输导作用 在植物体木质部和韧皮部中运输溶解物质的过程。

半透明的 允许光线穿过，但不透明的。

蒸腾作用 水分通过气孔散失到大气中。

横的 横向的。

梯形的 形似梯子的，即四边中只有两边平行的。

感伤性 植物对伤害响应的运动。

乔木 一类通常很高、具单一树干、其上着生由枝条组成的树冠的木本多年生植物。[162]

树冠 树林或森林中最高树的树冠形成的覆盖面。

年轮 （见"年轮"）。

三体雄蕊的 具有三组雄蕊的，如金丝桃属某些植物的花。

三雄蕊的 具三枚雄蕊的，如鸢尾科植物。

三角的 具三个角和三个面的。[173, 257]

三孔的 具三个萌发孔的。

三原型根 中柱内具三条原生木质部脊的根。[149]

族 通常为一个大的亚科下的一种次级等级，但有时可直接将科分为族。族的命名以"-eae"结尾。

trichome Any hair-like growth, glandular or eglandular, from the epidermis.

trichotomous Branching into three.

tricolpate Of a pollen-grain, having three colpi.

tricolporate Of a pollen grain, having three composite apertures, each consisting of a colpus and a pore.

tridentate With three teeth. **[195]**

trifid Divided to about half-way into three parts.

trifoliate Three-leaved.

trifoliolate With three leaflets. **[192]**

trigeneric Composed of three different genera, as the orchid x *Brassolaeliocattleya*, a hybrid genus produced by crossing a species of *Brassia* with one of *Laelia* and one of *Cattleya*.

trigger hairs Sensitive hairs, which when touched, cause a particular reaction. In e.g. *Dionaea muscipula* (Venus's Fly-trap), an insect touching one of the trigger hairs on the inner surface of the leaf will cause the two lobes to close rapidly, trapping the prey inside. Special glands then secrete digestive enzymes which act on the body of the insect, producing products from which the plant is able to obtain nutrients. **[202]**

trigonous Three-angled.

trijugate Of a compound leaf, having three pairs of leaflets. **[248]**

trilobate, trilobed With three lobes.

trilocular Having three loculi or compartments. **[252]**

trimerous Having the parts of the flower in threes, as in the Iridaceae.

trimorphic Occurring in three forms.

trioecious Having male, female, and bisexual flowers on different plants of the same species.

tripartite Divided almost to the base into three parts.

毛状体 任一从表皮上生出的具腺体的或无腺体的毛状生长物。

三歧式的 分支成三的。

三沟的 花粉粒具三个萌发沟的。

三孔沟的 花粉粒具三个复合孔的，每孔包括一个萌发沟和一个萌发孔。

三齿的 具有三枚齿的。[195]

三裂的 分裂至约一半成三部分。

具三叶的 具三枚叶的。

具三小叶的 具有三枚小叶的。[192]

三属的（杂交种） 由三个不同属组成的，像兰花×Brassolaeliocattleya，是一个由蜘蛛兰属（*Brassia*）的种与蕾丽亚兰属（*Laelia*）的种及嘉德利亚兰属（*Cattleya*）的种杂交而产生的杂交属。

触发柔毛 当被触动时引起特殊反应的敏感柔毛。如在捕蝇草中，触动叶内面上的触发柔毛之一的昆虫将导致两个裂片快速闭合，将其夹于里面。然后特殊腺体分泌消化酶作用于昆虫身体上，产生植物能从其获取营养的产物。[202]

三棱的 具三个棱的。

具三对小叶的 复叶具三对小叶的。[248]

三裂片的 具有三个裂片的。

三室的 三个室或腔的。[252]

三基数的 花各部分以三为基数的，像鸢尾科植物。

三形的 以三种形式出现的。

单全异株 同种植物不同植株上具有雄性、雌性和两性花的。

三深裂的 几乎分裂到基部成三部分的。

tripinnate Bipinnate, with the secondary leaflets again pinnate. **[192]**

三回羽状的　二回羽状的，具又羽状的次生小叶。[192]

tripinnatifid Literally bipinnatifid, with the secondary leaflets again pinnatifid, but frequently the primary division is pinnate. **[276]**

三回羽状半裂的　二回羽状半裂的，具又羽状半裂的次生小叶，但初级分裂往往也为羽状。[276]

triplinerved Having three main veins.

离基三出脉的　具有三条主脉的。

triploid (3n) Having three sets of chromosomes in each cell.

三倍体（3n）的　每个细胞内具三套染色体的。

triporate Of a pollen grain, having three pores.

三孔的　花粉粒具三个（萌发）孔的。

triquetrous Triangular in cross-section, with sharp angles sometimes produced by concave sides. **[246]**

三棱的　横截面呈三角形的，有时具由凹面产生的锐角。[246]

triradiate Having three rays. **[282]**

三射的　具三条射线的。[282]

tristyly The occurrence of three different lengths of style in flowers of the same plant species, as *Lythrum salicaria* (Purple Loosestrife).

三型花柱　同种植物花中具三种不同长度花柱的现象，像千屈菜。

trisulcate Of a pollen grain, having three grooves or furrows.

具三槽的　花粉粒具三条沟的。

tritegmic Of an ovule, having three integuments.

三珠被的　胚珠具三层珠被的。

tropic, tropical Occurring in the tropics, the region extending to about 23 degrees on either side of the equator.

热带的　发生在热带的，该地区延伸到赤道两侧约23度。

tropism A directional response by a plant to a stimulus, positive if towards the source of the stimulus or negative if away from it.

向性，趋性　植物对刺激物的方向性响应，朝向刺激源者为正响应，远离刺激源者为负响应。

tropophyte A plant adapted to a seasonal climate, and surviving periods unfavourable for growth by forming resting buds.

湿旱生植物　一类以形成休眠芽度过不利于生长时期的适应季节性气候的植物。

trullate Trowel-shaped. **[190]**

镘形　镘形的。[190]

trumpet-shaped Narrowly tubular, ending in a flared limb. **[211]**

喇叭形　狭管状，末端呈一展开的檐部。[211]

truncate Appearing as if cut off at the base or apex. **[194, 195]**

截形的，平截的　形似在基部或顶端突然切下的。[194, 195]

trunk The bole, or upright main stem of a tree which bears a crown of branches. **[162, 272]**

树干　树干或树木的直立主干，生有由树枝组成的树冠。[162, 272]

truss A cluster of flowers or fruit growing on a single stalk.

花束，果束　生于单个柄上的一束花或果实

T.S. Transverse section.

横断面，横切面。

tube nucleus The nucleus of the vegetative cell, the larger of the two cells into which the nucleus of the pollen grain divides while still in the pollen sac. **[213]** (see also **generative cell**)

tuber An underground stem or root, swollen with reserves of food. **[152]**

tubercle A small tuber; a small, rounded projection. **[220, 240]**

tubercled, tuberculate (see **verrucose**).

tuberous Resembling or producing tubers.

tubular Cylindrical and hollow. **[211]**

tubular floret A regular floret, usually a disc floret in the Compositae. **[234]**

tunic The outer, dry, papery covering of a bulb or corm, often fibrous or reticulate. **[151]**

tunica The outer layer or layers of cells in an apical meristem, which divide to produce the epidermis of the shoot.

tunicate Consisting of a number of concentric layers, the outermost usually dry and membranous, as the bulb of *Allium cepa* (Onion). **[151]**

turbinate (obconical) Top-shaped, inversely conical. **[160]**

turgid Of cells, swollen and rigid as a result of the uptake of water.

turgor Rigidity of cells resulting from their uptake of water.

turion An underground bud or shoot which develops into an aerial stem; a winter bud, characteristic of many aquatic plants, that contains food material and becomes detached from the parent plant, either floating or resting at the bottom of the water until favourable conditions stimulate its development and growth into a new plant, as in *Potamageton crispus*. **[154, 166]**

tussock A clump or tuft, especially of a grass.

twig One of the smallest divisions of the branch of a tree. **[170]**

粉管核 营养细胞核，仍在花粉囊中时，花粉粒核即分裂成两个细胞，其中较大的细胞，即营养细胞。[213]（又见"生殖细胞"）

块茎，块根 一类随食物营养的储存而膨大的地下茎或根。[152]

小瘤，突起 小块茎状突起；小圆形突起。[220, 240]

具小瘤的 （见"疣状的"）。

块茎状的 类似或产生块茎的。

管状的，筒状的 圆柱状且空心的。[211]

管状小花 一类整齐花，菊科植物通常为盘花。[234]

鳞茎皮 鳞茎或球茎外面的干燥纸质、通常为纤维状或网状的外层覆盖物。[151]

原套 能够分裂产生嫩枝表皮的、顶端分生组织的最外层细胞。

具膜被的 由大量的同心层组成，其最外层通常干燥、膜质，如洋葱的鳞茎。[151]

陀螺状的 陀螺形的，倒锥形的。[160]

膨胀的 细胞由于吸水而膨胀僵硬的。

膨胀 由于细胞吸水而变得僵硬。

具鳞根出条、冬芽 一类能够发育成气生茎的地下芽或嫩枝；一种含有食物营养的、从母株上分离下来的、漂浮或停留水底直到有利条件刺激其生长发育成一新植株的冬芽，是许多水生植物的特征，如菹草。[154, 166]

草丛 一丛或一簇，特别是禾草类植物。

小枝 树枝的最小分枝之一。[170]

tylose, tylosis In woody plants, an outgrowth of the wall of a parenchyma cell which protrudes into an adjacent duct.

侵填体 木本植物（茎内）薄壁组织细胞壁侵入相邻导管内（形成）的生长物。

uliginous Growing in swamps.

湿地的 沼泽中生长的。

umbel An inflorescence in which the pedicels arise from the same point on the peduncle. An umbel can range from being flat-topped, as in *Daucus carota* (Carrot) to almost spherical, as in *Allium cepa* (Onion). **[205]**

伞形花序 一类从花序柄同一点生出花梗的花序，伞形花序从如胡萝卜花一样的平顶，到洋葱花一样的球面。[205]

umbellate In the form of an umbel; bearing umbels.

伞形的，具伞形花序的 以伞形花序的形式；生有伞形花序的。

umbellifer A member of the Umbelliferae.

伞形植物 伞形科植物。

umbelliferous Bearing an umbel or umbels, typically a member of the Umbelliferae.

具伞形花序的 生有伞形花序或复伞形花序的，典型的如伞形科植物。

umbelliform In the form of an umbel.

伞形花序状的 以伞形花序形式的。

umbo A boss or protuberance, especially that which occurs on a cone scale. **[270]**

鳞脐 一种通常在果鳞上生出的尖头或突起。[270]

umbraculate Parasol-shaped. **[225]**

伞形的 太阳伞状的。[225]

unarmed Without spines, prickles or thorns.

无刺的 没有刺、皮刺或枝刺的。

uncinate Hooked.

具钩的 具钩的。

undershrub (see **subshrub**).

半灌木 （见"亚灌木"）。

understorey A layer of vegetation composed of shrubs and small trees lying between the tree canopy and the ground cover in a wood or forest.

下层林木 在树林或森林中，位于乔木树冠和地被层之间、由灌木和小乔木组成的植被层。

undulate With a wavy margin curving up and down. **[193]**

波状的 具上下弯曲的波状边缘的。[193]

unguiculate Narrowed at the base into a claw. **[227]**

具爪的 基部变狭成爪的。[227]

uniaperturate Having one pore.

单孔的 具一个（萌发）孔的。

unicarpellate Having one carpel.

单心皮的 具一枚心皮的。

unicellular One-celled. **[202]**

单细胞的 具一个细胞的。[202]

unijugate Of a compound leaf, having one pair of leaflets as in *Lathyrus odoratus* (Sweet Pea). **[177]**

具一对小叶的 像香豌豆一样，复叶具有一对小叶的。[177]

unilateral One-sided.

单侧的 一侧的。

unilocular Having a single loculus or compartment. **[219]**

单室的 具有单个腔或室的。[219]

uniovulate Having a single ovule.

单胚珠的 具有单个胚珠的。

uniseriate Arranged in a single row or series.

单列的 排成单一的行或单一的列的。

[265]

unisexual (diclinous) Having only male or female organs in the flower. The male and female flowers may be on separate plants (dioicism) or on the same individual (monoecism). **[227]**

united Joined. **[207, 225]**

unitegmic Of an ovule, having one integument.

upright Erect; with the branches growing upwards. **[163]**

urceolate Urn-shaped, pitcher-shaped. **[211]**

utricle A bladder-like structure, especially the membranous sac in the fruit of the genus *Carex* (sedge) the fruit in members of Cyperaceae. **[225, 246]**

vaginate Sheathed. **[187]**

vallecular canal Any of the canals in the stem of species of the genus *Equisetum* (horsetail), each one opposite a groove on the surface. **[283]**

valvate When similar parts of the plant meet exactly without overlapping **[170]**; in anthers or fruits, opening by valves. **[216, 256]**

valve One of the pieces into which an anther splits at maturity to release the pollen; one of the pieces into which a fruit splits at maturity to release the seeds. **[256]**

valvular dehiscence Splitting open by means of valves. **[255, 256]**

var. Variety.

variation Divergence from the normal state, usually due to genetic differences or influence of the environment.

variety A rank used to designate a group of plants varying in flower colour, habit, or some other way.

vars. Varieties.

variegated Having two or more colours in the leaves.

[265]

单性的（雌雄异花的） 花上只有雄性或雌性器官的。雄花和雌花可位于不同植株上或同一植株上。[227]

联生的 连接的。[207, 225]

单珠被的 胚珠具一层珠被的。

直立的 垂直的；具有向上生长的枝的。[163]

坛状的，罐状的。[211]

胞囊，胞果 一类囊状结构特别是苔草属植物果实的膜状囊（莎草科植物的果实类型）。[225, 246]

具鞘的 有鞘的。[187]

槽腔 木贼属植物茎中的任一腔道，每一腔道对应着表面一条沟。[283]

镊合状的，瓣裂的 植物体的相同部分相互接触，但不覆盖 [170]；在花药或果实中以瓣片开裂。[216, 256]

裂爿 花药成熟时以其开裂散出花粉的或果实成熟时以其开裂散出种子的瓣片之一。[256]

瓣裂 以瓣裂的方式裂开。[255, 256]

变种 变种（单数）。

变异 通常由于遗传上不同或环境影响产生的不同于正常状态的差异。

变种 一种用来表明一群植物在花色、习性或其他方面变异的分类等级。

变种 变种（复数）。

杂斑的 叶片上具有两种或多种颜色的。

vascular Relating to the vessels and sieve tube that convey water and nutrients within the plant, i.e. the xylem and phloem.

vascular bundle One of the strands of tissue that conduct water and nutrients within the plant, consisting of xylem on the inside and phloem on the outside, separated by a layer of cambium. **[149, 168—171, 173, 196, 212, 219]**

vascular cambium The layer of meristematic cells lying between xylem and phloem that produces additional xylem and phloem, resulting in the lateral growth of a stem or root. **[169, 171]**

vascular cryptogam A cryptogam that resembles flowering plants in having a vascular system, as a fern or a fern ally. **[275—284]**

vascular cylinder (see **stele**).

vascular plants (Tracheophyta, tracheophytes) Plants that have a vascalar system of xylem and phloem to conduct water and nutrients, i.e. members of the Pteridophyta and Spermatophyta.

vascular ray One of the secondary medullary rays, formed by the vascalar cambium, that supplement the primary medullary rays.

vascular strand (see **vascular bundle**).

vascular system (vascular tissue) The conducting tissue in a vascular plant, consisting of xylem and phloem, separated by a layer of cambium. **[148]**

vegetative Relating to the non-flowering parts of a plant.

vegetative apomixis A form of apomixis in which plants reproduce vegetatively, by rhizomes, stolons, runners, stem tubers, or bulbils.

vegetative propagation Asexual reproduction, i.e. by means of bulbs, rhizomes, runners etc. rather than by seed. **[154, 155]**

vein (nerye) A strand of vascular tissue in a

维管的 与植物体内输导水和养料的导管（和筛管）相关的，即木质部和韧皮部。

维管束 一类在植物体内输导水分和养料的组织束之一，包括被形成层分开的内侧木质部和外侧韧皮部。 [149, 168—171, 173, 196, 212, 219]

维管形成层 一层位于木质部和韧皮部之间、产生次生木质部和次生韧皮部、导致茎或根侧面生长（或加粗）的分生组织细胞层。 [169, 171]

维管隐花植物 一类像蕨类和拟蕨类一样、在具维管系统方面类似于有花植物的隐花植物。 [275—284]

维管柱 （见"中柱"）。

维管植物（导管植物门，维管植物） 一类由木质部和韧皮部组成的维管系统来运输水分和养料的植物，即蕨类植物和种子植物。

维管射线 一类由维管形成层产生、补充初生髓质射线的次生髓质射线。

维管束 （见"维管束"）。

维管系统（维管组织） 维管植物由被形成层分开的木质部和韧皮部组成的运输导组织。 [148]

营养体的 与植物非开花部分相关的。

营养体无性生殖 植物通过根状茎、匍匐茎、块茎或珠芽进行营养繁殖的一种无性生殖方式。

无性繁殖 无性生殖，即以鳞茎、根状茎、匍匐茎等繁殖，而不是以种子繁殖。 [154, 155]

脉 叶或其他扁平器官中的维管组织束。

leaf or other flat organ. **[168, 196]**

veinlet A small vein. **[278]**

velamen The outer layer of the aerial roots of epiphytic orchids and aroids, consisting of thick-walled, non-living cells. **[148]**

velum The membranous indusium in the genus *Isoetes* (quillwort). **[281]**

velutinous Velvety. **[200]**

venation The arrangement of veins, as in a leaf. **[189]**

ventral The side of an organ facing towards the axis, adaxial. **[216, 256]**

ventricose Swollen or inflated on one side, as the corolla in some members of the Labiatae and Scrophulariaceae.

ventriculose Slightly ventricose.

vernacular name The common name for a plant in any language, as opposed to the scientific name.

vernal Occurring in spring.

vernation The arrangement of leaves in a vegetative bud. **[189]** (see also ptyxis)

verrucose (tubercled, tuberculate) Bearing small, wart-like projections. **[199, 220]**

versatile With the anther attached at the middle and turning freely on its filament. [216]

verticil (whorl) The arrangement of three or more organs in a circle round the axis. **[186]**

verticillaster (false whorl) A deceptive kind of inflorescence found in the Labiatae, which gives the appearance of a whorl but which in reality consists of two dichasial cymes on opposite sides of the stem. **[205]**

verticillate Arranged in a whorl. **[186]**

vesicle A small bladder or sac.

vesiculose Covered with small bladders. **[199]**

vespertine Of flowers, opening in the evening.

vessel In angiosperms, one of the tubular structures forming the xylem and phloem

[168, 196]

细脉 小叶脉。 [278]

根被 附生兰花和天南星科植物气生根、由厚壁死细胞组成的最外层。 [148]

缘膜 水韭属植物的膜质舌片。 [281]

被短绒毛的 被茸毛的。 [200]

脉序 像叶中叶脉的排列方式。 [189]

腹面的 面向轴的器官面，近轴面的。 [216, 256]

一面臌的 一面肿胀或膨胀，如唇形科和玄参科植物的花。

一面稍臌的 一面稍膨大。

俗名 任一语言的植物通用名称，相对于学名。

春季开花的 春季发生的。

幼叶卷叠式 叶子在芽内的排列方式。 [189]（又见"小叶卷叠式"）

疣状的（瘤状的，小瘤状的） 具小疣状突起的。 [199, 220]

丁字着的 花药以中部着生于花丝顶端并可以自由转动的。 [216]

轮 三个或多个器官围绕轴着生为一圈的排列方式。 [186]

轮状聚伞花序，轮伞花序（假轮生） 一类在唇形科植物中发现的假花序，从外表看为轮生，但实际上包括两个生于茎相对两侧的二歧聚伞花序。 [205]

轮生的 呈一轮着生的。 [186]

小泡，泡囊 一类小泡或小囊。

多小泡的，囊状的或泡状的 覆盖有小泡的或小囊的。 [199]

暮开性的 花在夜晚开放的。

导管 在被子植物中，一类形成木质部（原著"和韧皮部"）的、运输水分（原著

that conduct water and nutrients throughout the plant.

vestigial Imperfect development of an organ which was fully developed in some ancestral form.

vestiture A covering of hairs or scales. [199—202]

vexillary (see **imbricate-descending**).

vexillum The standard, the large upper petal in the flowers of plants of subfamily Papilionoideae in the Leguminosae. [230, 255]

viable Of seeds, able to germinate.

vicariad, vicariant One of two similar taxa occupying separate geographical areas, e.g. *Jasminum nudiflorum* or *J. mesnyi*.

villose, villous Covered with long, shaggy hairs. [201]

vine A woody or herbaceous plant with a long, climbing, scrambling, or trailing stem, especially one belonging to the grape family (Vitaceae).

virgate Long, straight, and slender.

virgulate Diminutive of virgate.

viscid (viscous) Sticky.

viscidium (viscid disc) The sticky disc at the base of the caudicle in the Orchidaceae which adheres to the head of a visiting insect. [252]

viscin A sticky substance surrounding the seeds in members of the Loranthaceae.

viscous (see **viscid**).

vitta An oil-canal in the fruits of the Umbelliferae. [257]

viviparous Reproducing by vivipary. [155]

vivipary The production of buds that form plantlets while still attached to the parent plant; the production of seeds that germinate within the fruit. [155]

volute Rolled up.

"和养料"）和无机盐至植物全身的管状结构之一。

残迹的，废退的 一类祖先型完全发育的器官的不完全发育状态。

表被 一类由毛或鳞片构成的覆盖物。[198—202]

蝶形卷叠式 （见"下向覆瓦状的"）。

旗瓣 豆科蝶形花亚科植物花的大的上面花瓣。[230, 255]

能发芽的种子 种子能萌发的。

替代种 占据不同地理区域的两个相似类群之一，如迎春花或野迎春。

具长柔毛的 覆盖有长而蓬松柔毛的。[201]

藤本植物 一类具细长、爬行、攀缘或蔓延茎的木本或草本植物，尤其是属于葡萄科的植物。

多直细枝的，帚状的 长的、直的、细的。

小帚状的 小帚状的。

黏性的 黏性的。

黏盘 兰科植物花粉块柄基部的可以黏附于访问昆虫头部的黏性盘状物。[252]

黏滞的，黏的 包围在桑寄生科植物种子周围的黏性物质。

黏滞的 （见"黏性的"）。

油道 一类存在于伞形科植物果实内的油道。[257]

胎生的 通过胎生方式生殖的。[155]

胎生，胎萌 产生附着于亲本植株、能够形成小植株的芽的繁殖方式；产生在果实内萌发的种子的繁殖方式。[155]

向上卷的 卷起来的。

weeping With the branches bending over and hanging down. **[253]**

whorl (see **verticil**).

whorled (see **verticillate**).

window, window pane (see **fenestration**).

wing A flat, often dry and membraneous extension to an organ **[255, 264]**; one of the lateral petals (alae) in the flowers of plants of subfamily Papilionoideae in the Leguminosae. **[230]**

winged (see **alate**).

winter annual An annual plant that germinates in the autumn, and survives the winter as a rosette of leaves.

witches' broom An abnormally dense tuft of twigs in a tree, resulting from an attack by fungi, mites, or viruses. **183]**

wood The secondary xylem of trees; the constructional material obtained from felled tree. **[170]** (see **softwood** and **hardwood**)

x Before a botanical plant name, indicates a sexual hybrid; before a number, indicates the degree of enlargement.

xanthophylls A group of yellow pigments, belonging to the carotenoids, that occur in the chromoplasts of plant cells.

xenogamy A form of allogamy, in which the ovules of a flower are fertilised by pollen from a flower on a different plant.

xerad (see **xerophyte**).

xeromorphic Having the characteristics of plants growing in dry places, i.e. reduced or succulent leaves and stems, dense hairiness or a thick cuticle.

xerophyte (xerad) A plant that is adapted to grow in a dry habitat. **[241]**

xerophytic Growing in dry habitats. **[241]**

xylem The vascular tissue that conducts water and minerals from the roots to other parts of the plant. **[148, 168]**

下垂的 树枝弯曲、垂下来的。[253]

轮 （见"轮"）。

轮生的 （见"轮生的"）。

窗，窗格 （见"窗口"）。

翅，翼瓣 器官的扁平、干燥、膜质延伸物 [255, 264]；豆科蝶形花亚科植物花的侧生花瓣（翼瓣）之一。[230]

具翅的 （见"具翅的"）。

冬性一年生植物 在秋季萌发，以莲座叶丛艰难度过冬季的一年生植物。

丛枝病 一棵树上由真菌、螨虫或病毒攻击导致的异常密集的细枝丛。[183]

木材 树木的次生木质部；从砍伐的树木获得的建筑材料。[170]（见"软木材"和"硬木材"）

杂种的表示符号，或放倍数 在植物名称之前，表示有性杂种；在数字之前，表示放大倍数。

叶黄素 一类植物细胞有色体中出现的、属于类胡萝卜素的黄色色素。

异株 异花受精，异花受精的一种形式。即一朵花上的胚珠被来自不同植株花上的花粉受精。

旱生植被 （见"旱生植物"）。

旱生结构的 具生长在干旱地区的植物的特征，即退化或肉质的叶和茎，密集的毛被或厚的角质层。

旱生植被（旱生植被） 适于生长在干旱生境的植物。[241]

旱生性的 在干旱生境生长的。[241]

木质部 从根到植物体其他部分输送水分和矿质营养的维管组织。[148, 168]

xylocarp A haid, woody fruit.

zygomorphic (bilaterally symmetric) Divisible through the centre of the flower in only one longitudinal plane for the halves of the flower to be mirror images, as many members of the Labiatae and Scrophulariaceae. **[209, 226]**

zygote The cell (usually diploid) that results from the fusion of a male and a female gamete.

zygotic Relating to a zygote.

硬木质果 坚硬、木质的果实。

两侧对称的 只有一个纵向面可通过花的中心，将花分为互为镜像的两部分，如唇形科和玄参科的多数植物。 [209, 226]

合子 由雌雄配子融合产生的细胞（通常为二倍体）。

合子的 与合子有关的。

Illustrations
图解

1. Roots, Storage Organs and Vegetative Reproduction
1. 根、贮藏器官和营养繁殖

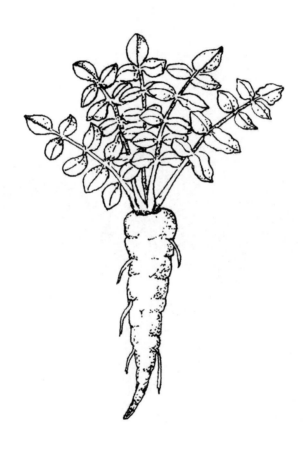

Roots 1
根 1

Fibrous roots of *Cerastium* sp.
卷耳属植物的须根

Abventitious roots of a grass
禾本科植物的不定根

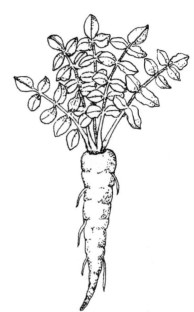

Taproot of *Pastinaca sativa*
欧防风的直根

Root tubers of *Ranunculus ficaria*
榕叶毛茛的块根

Root tubers of *Ophrys* sp.
眉兰属植物的块根

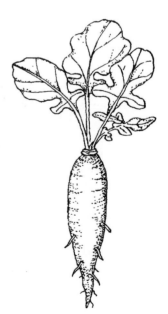

Taproot of *Raphanus sativus*
萝卜的直根

Roots 2
根 2

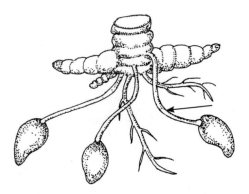

Nodulose roots of *Curcuma amada*
芒果姜的具瘤根

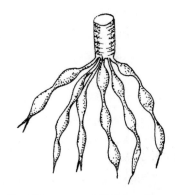

Moniliform roots of *Momordica* sp.
苦瓜属植物的念珠状根

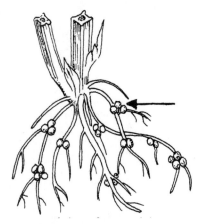

Root nodules of *Vicia faba*
蚕豆的根瘤

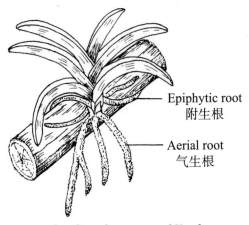

Epiphytic root
附生根

Aerial root
气生根

Aerial and epiphytic roots of *Vanda* sp.
万代兰属植物的气生根和附生根

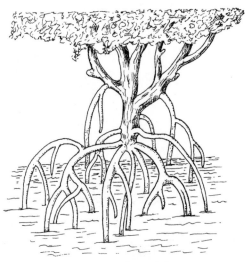

Prop roots of *Rhizophora* sp.
红树属植物的支柱根

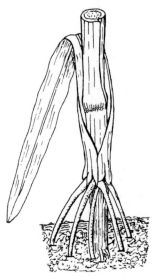

Prop roots of *Zea mays*
玉米的支柱根

Roots 3
根 3

Pneumatophores of *Taxodium distichum*
落羽杉的出水通气根

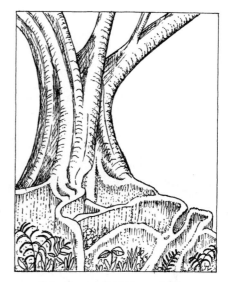

Buttress roots of *Ficus elastica*
橡皮树的板状根

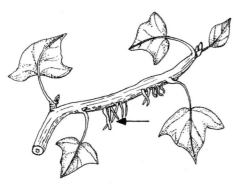

Adventitious clinging roots of *Hedera helix*
洋常春藤的攀缘不定根

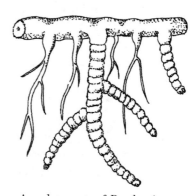

Annulate roots of *Psychotria* sp.
九节属植物的环节根

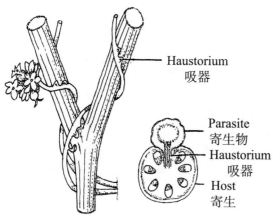

Haustorium
吸器

Parasite
寄生物
Haustorium
吸器
Host
寄生

Haustoria of *Cuscuta* sp.
菟丝子属植物的吸器

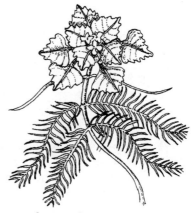

Plumose photosynthetic roots of *Trapa natans*
菱角的羽状光合根

Roots 4
根 4

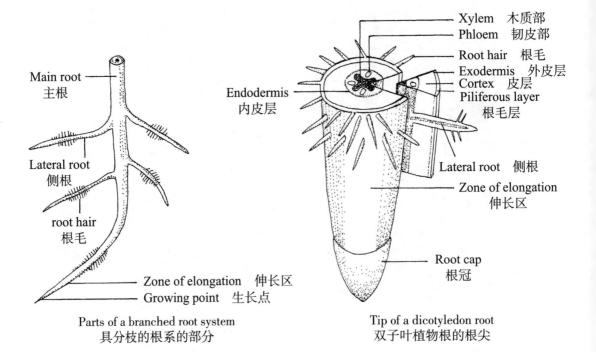

Main root
主根

Lateral root
侧根

root hair
根毛

Zone of elongation　伸长区
Growing point　生长点

Parts of a branched root system
具分枝的根系的部分

Xylem　木质部
Phloem　韧皮部
Root hair　根毛
Exodermis　外皮层
Cortex　皮层
Piliferous layer
根毛层

Endodermis
内皮层

Lateral root　侧根

Zone of elongation
伸长区

Root cap
根冠

Tip of a dicotyledon root
双子叶植物根的根尖

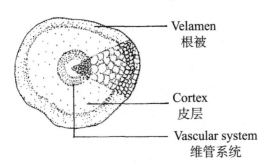

Velamen
根被

Cortex
皮层

Vascular system
维管系统

T.S. of an epiphytic root of *Vanda* sp.
万代兰属植物的附生根横切

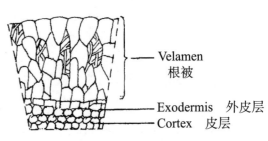

Velamen
根被

Exodermis　外皮层
Cortex　皮层

Detail of the velamen
根被详图

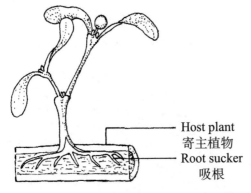

Host plant
寄主植物
Root sucker
吸根

Root suckers of the hemi-parasite *Viscum album*
半寄生植物槲寄生的吸根

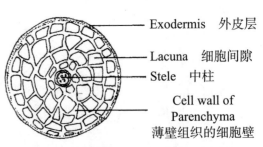

Exodermis　外皮层

Lacuna　细胞间隙

Stele　中柱

Cell wall of
Parenchyma
薄壁组织的细胞壁

T.S. of stem of *Hippuris vulgaris* showing lacunae
杉叶藻茎的横切，示细胞间隙

Roots 5
根 5

DIAGRAMS OF ROOT TYPES　根类型示意图

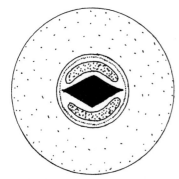

T.S. of diarch root
found in many ferns and gymnosperms
二原型根横切
在许多蕨类和裸子植物中发现

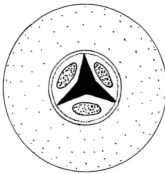

T.S. of triarch root
found in *Pinus* spp.
三原型根横切
在松属中发现

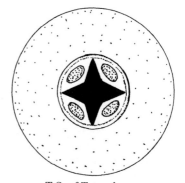

T.S. of Tetrach root
found in many dicotyledons
四原型根横切
在许多双子叶植物中发现

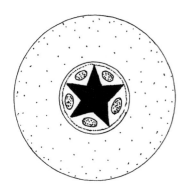

T.S. of pentarch root
found in some dicotyledons
五原型根横切
在某些双子叶植物中发现

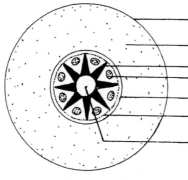

Piliferous layer 根毛层
Cortex　皮层
Xylem　木质部 ⎫ Vascular
Phloem　韧皮部 ⎬ bundles
Endodermis　内皮层 ⎭ 维管束
Pericycle　中柱鞘
Pith (present in some cases only, e.g. *Zea mays*)
髓（仅在某些情况下存在，如玉米）

T.S. of polyarch root
found in all monocotyledons
多原型根横切
在所有单子叶植物中发现

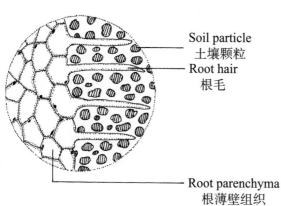

Soil particle
土壤颗粒
Root hair
根毛
Root parenchyma
根薄壁组织

Detailed drawing of root hairs
根毛详图

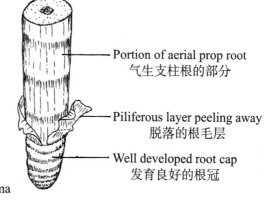

Portion of aerial prop root
气生支柱根的部分
Piliferous layer peeling away
脱落的根毛层
Well developed root cap
发育良好的根冠

Tip of an aerial root of *Pandanus nobilis*
菲律宾露兜树的气生根的根尖

Storage Organs 1
贮藏器官 1

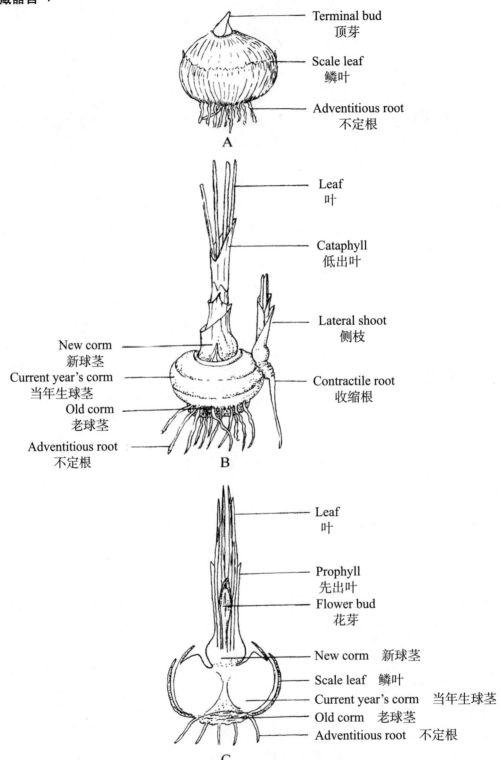

Terminal bud
顶芽

Scale leaf
鳞叶

Adventitious root
不定根

A

Leaf
叶

Cataphyll
低出叶

Lateral shoot
侧枝

New corm
新球茎

Current year's corm
当年生球茎

Old corm
老球茎

Contractile root
收缩根

Adventitious root
不定根

B

Leaf
叶

Prophyll
先出叶

Flower bud
花芽

New corm   新球茎

Scale leaf   鳞叶

Current year's corm   当年生球茎

Old corm   老球茎

Adventitious root   不定根

C

Corm of *Crocus* sp.   A. Dormant stage   B. Growing stage with scale leaves removed   C. L.S. of corm
番红花属某植物的球茎   A. 休眠期   B. 脱去鳞叶的生长期   C. 球茎纵切

Storage Organs 2
贮藏器官 2

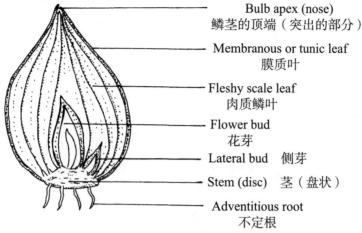

Bulb apex (nose)
鳞茎的顶端（突出的部分）

Membranous or tunic leaf
膜质叶

Fleshy scale leaf
肉质鳞叶

Flower bud
花芽

Lateral bud　侧芽

Stem (disc)　茎（盘状）

Adventitious root
不定根

L.S. of Onion, *Allium cepa*
洋葱纵切

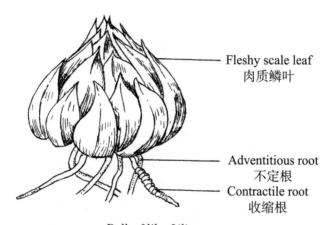

Fleshy scale leaf
肉质鳞叶

Adventitious root
不定根

Contractile root
收缩根

Bulb of lily, *Lilium* sp.
百合属植物的鳞茎

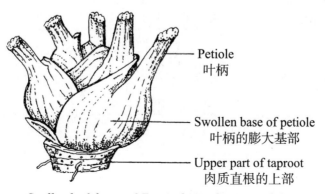

Petiole
叶柄

Swollen base of petiole
叶柄的膨大基部

Upper part of taproot
肉质直根的上部

Swollen leaf−bases of *Foeniculum vulgare* var. *dulce*
佛罗伦萨茴香的膨大叶基

Storage Organs 3
贮藏器官 3

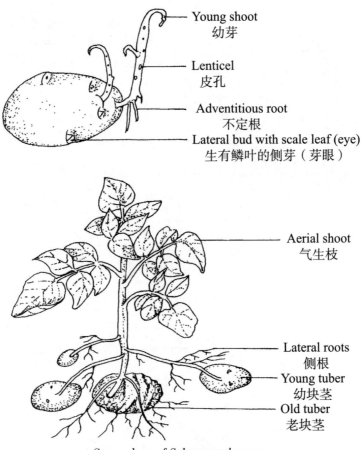

Young shoot
幼芽

Lenticel
皮孔

Adventitious root
不定根

Lateral bud with scale leaf (eye)
生有鳞叶的侧芽（芽眼）

Aerial shoot
气生枝

Lateral roots
侧根

Young tuber
幼块茎

Old tuber
老块茎

Stem tubers of *Solanum tuberosum*
马铃薯块茎

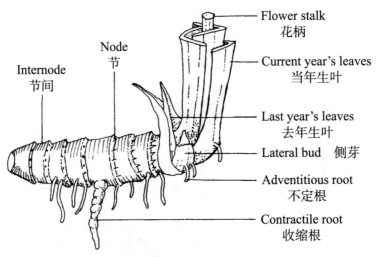

Flower stalk
花柄

Node
节

Current year's leaves
当年生叶

Internode
节间

Last year's leaves
去年生叶

Lateral bud   侧芽

Adventitious root
不定根

Contractile root
收缩根

Rhizome of *Iris* sp.
鸢尾属植物的根状茎

Storage Organs 4
贮藏器官 4

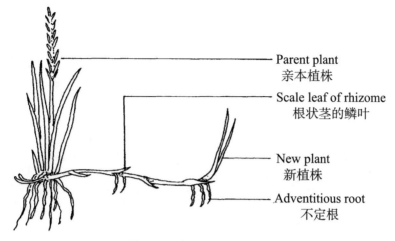

Parent plant
亲本植株

Scale leaf of rhizome
根状茎的鳞叶

New plant
新植株

Adventitious root
不定根

Rhizome of *Agropyron repens*
偃麦草的根状茎

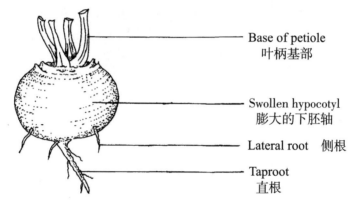

Base of petiole
叶柄基部

Swollen hypocotyl
膨大的下胚轴

Lateral root　侧根

Taproot
直根

Swollen hypocotyl of *Brassica rapa*
芜菁的膨大下胚轴

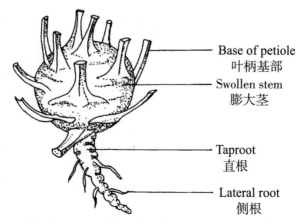

Base of petiole
叶柄基部

Swollen stem
膨大茎

Taproot
直根

Lateral root
侧根

Swollen stem of *Brassica oleracea* var. *gongylodes*
球茎甘蓝的膨大茎

Vegetative Reproduction 1
营养繁殖 1

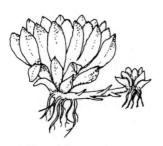

Offset of *Sempervivum* sp.
长生草属植物的短匍茎

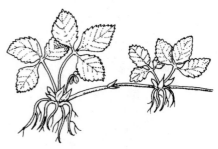

Runner of *Fragaria* sp.
草莓属植物的纤匍枝

Turions of *Hydrocharis morsus–ranae*
蟾蜍水鳖的冬芽

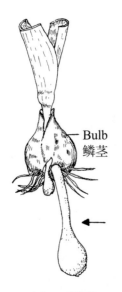

Dropper or sinker of *Tulipa* sp.
郁金香属植物的茎出条

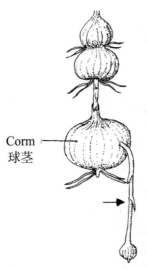

Dropper or sinker of *Crocosmia* sp.
雄黄兰属植物的茎出条

Sucker of *Mentha* sp.
薄荷属植物的地下匍匐茎

Aerial bulbils of *Allium sativum*
蒜的气生珠芽

Vegetative Reproduction 2
营养繁殖 2

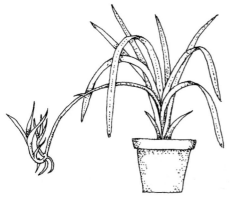

Plantlets borne on the long shoots of
Chlorophytum comosum
吊兰长枝上的小植株

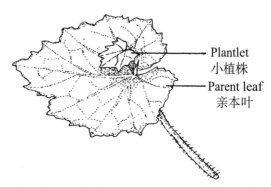

Plantlet
小植株

Parent leaf
亲本叶

Plantlet arising from junction of leaf blade and
petiole in *Tolmiea menziesii*
千母草叶片和叶柄连接处生出的小植株

A. Leaf of *Kalanchoe daigremontiana* with
plantlets (propagules)
生有小植株（繁殖体）的大叶落地生根的叶

B. Propagule
繁殖体

C. Rooted propagule
具根的繁殖体

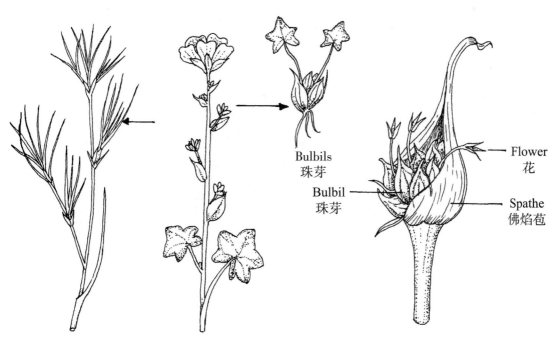

Bulbils
珠芽

Bulbil
珠芽

Flower
花

Spathe
佛焰苞

Vivipary in *Festuca vivipara*
胎生羊茅（新拟）的胎生
现象

Bulbils forming in place of lateral
flowers in *Saxifraga cernua*
零余虎耳草侧花位置形成的珠芽

Aerial bulbils in *Allium sativum*
蒜的气生珠芽

2. Seeds and Seedlings
2. 种子和幼苗

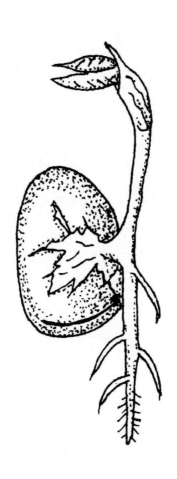

Seeds and Seedlings 1
种子和幼苗 1

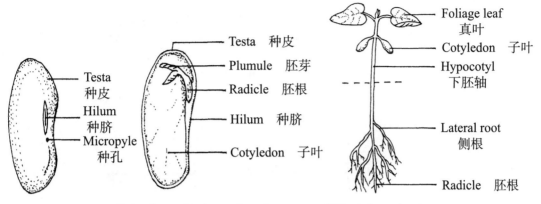

Epigeal germination and seed structure of *Phaseolus vulgaris*
菜豆的出土萌发和种子结构

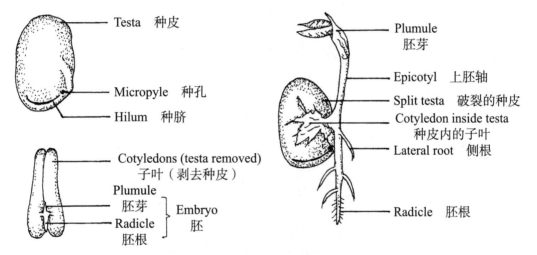

Hypogeal germination and seed structure of *Vicia faba*
蚕豆的留土萌发和种子结构

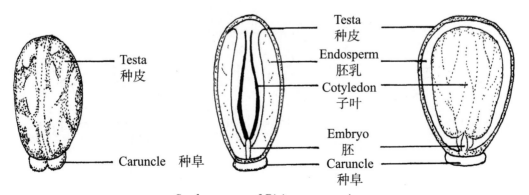

Seed structure of *Ricinus communis*
蓖麻的种子结构

Seeds and Seedlings 2
种子和幼苗 2

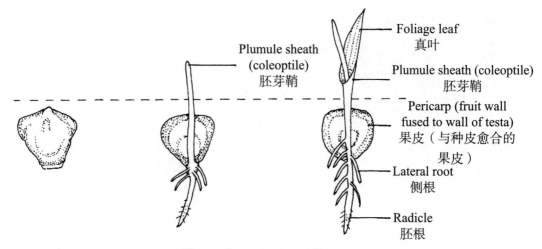

Foliage leaf
真叶

Plumule sheath (coleoptile)
胚芽鞘

Plumule sheath (coleoptile)
胚芽鞘

Pericarp (fruit wall fused to wall of testa)
果皮（与种皮愈合的果皮）

Lateral root
侧根

Radicle
胚根

Hypogeal germination of *Zea mays*
玉米的留土萌发

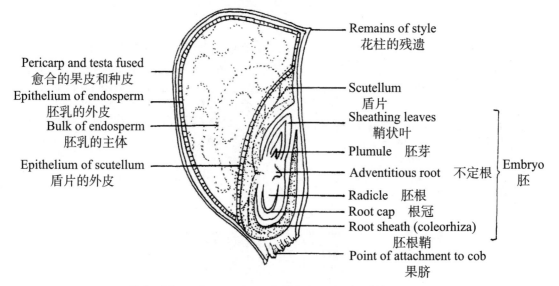

Remains of style
花柱的残遗

Pericarp and testa fused
愈合的果皮和种皮

Epithelium of endosperm
胚乳的外皮

Bulk of endosperm
胚乳的主体

Epithelium of scutellum
盾片的外皮

Scutellum
盾片

Sheathing leaves
鞘状叶

Plumule 胚芽

Adventitious root 不定根

Radicle 胚根

Root cap 根冠

Root sheath (coleorhiza)
胚根鞘

Point of attachment to cob
果脐

Embryo
胚

L.S. of the endospermic seed in the caryopsis of *Zea mays*
玉米颖果具胚乳种子的纵切

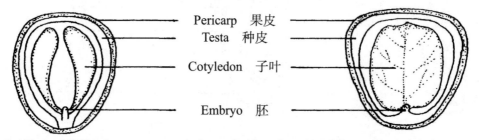

Pericarp 果皮
Testa 种皮

Cotyledon 子叶

Embryo 胚

L.S. of fruit of *Helianthus annuus*, cut in 2 ways to show the testa of the seed joined to the pericarp
向日葵果实的纵切，两种切法，显示结合到果皮上的种皮

Seeds and Seedlings 3
种子和幼苗 3

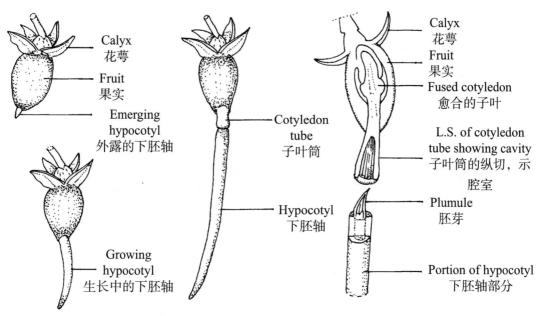

Calyx
花萼

Fruit
果实

Emerging
hypocotyl
外露的下胚轴

Growing
hypocotyl
生长中的下胚轴

Cotyledon
tube
子叶筒

Hypocotyl
下胚轴

Calyx
花萼

Fruit
果实

Fused cotyledon
愈合的子叶

L.S. of cotyledon
tube showing cavity
子叶筒的纵切，示
腔室

Plumule
胚芽

Portion of hypocotyl
下胚轴部分

Stages in germination of *Rhizophora mangle*
大红树萌发期

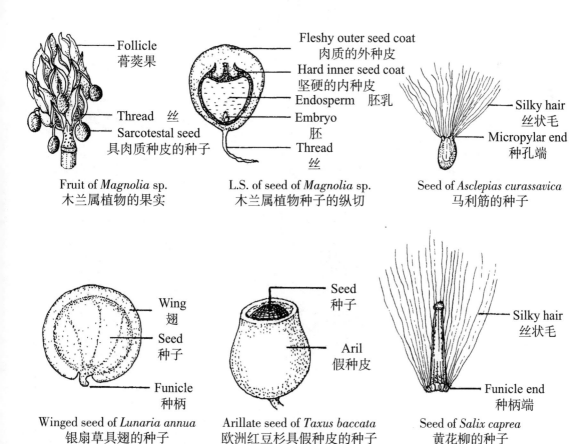

Follicle
蓇葖果

Thread 丝
Sarcotestal seed
具肉质种皮的种子

Fruit of *Magnolia* sp.
木兰属植物的果实

Fleshy outer seed coat
肉质的外种皮
Hard inner seed coat
坚硬的内种皮
Endosperm 胚乳
Embryo
胚
Thread
丝

L.S. of seed of *Magnolia* sp.
木兰属植物种子的纵切

Silky hair
丝状毛
Micropylar end
种孔端

Seed of *Asclepias curassavica*
马利筋的种子

Wing
翅
Seed
种子

Funicle
种柄

Winged seed of *Lunaria annua*
银扇草具翅的种子

Seed
种子

Aril
假种皮

Arillate seed of *Taxus baccata*
欧洲红豆杉具假种皮的种子

Silky hair
丝状毛

Funicle end
种柄端

Seed of *Salix caprea*
黄花柳的种子

Seeds and Seedlings 4
种子和幼苗 4

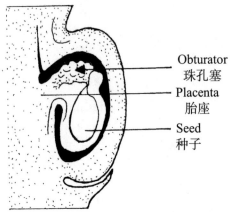

Obturator
珠孔塞
Placenta
胎座
Seed
种子

L.S. of portion of fruit of
Ricinus communis
蓖麻果实的部分纵切面

Cotton boll at dehiscence
开裂的棉铃

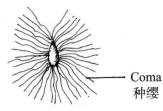

Coma
种缨

Seed with coma
具种缨的种子

Fruit and seed of *Gossypium* sp.
棉属植物的果实和种子

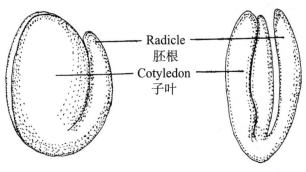

Radicle
胚根
Cotyledon
子叶

Accumbent cotyledons of
Rorippa nasturtium–aquaticum
豆瓣菜的缘倚子叶

Incumbent cotyledons of
Capsella bursa-pastoris
荠菜的背倚子叶

Lenticular seed of *Amaranthus* sp.
苋属植物的透镜状种子

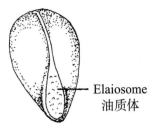

Elaiosome
油质体

Seed of *Viola reichenbachiana*
林生堇菜的种子

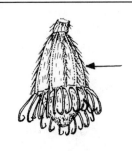

Turbinate calyx tube of
Agrimonia eupatoria
欧洲龙芽草陀螺状的萼筒

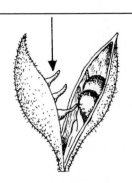

Jaculators in *Ruellia caroliniana*
卡罗莱纳蓝花草的种柄钩

3. Growth and Life Forms
3. 生长型和生活型

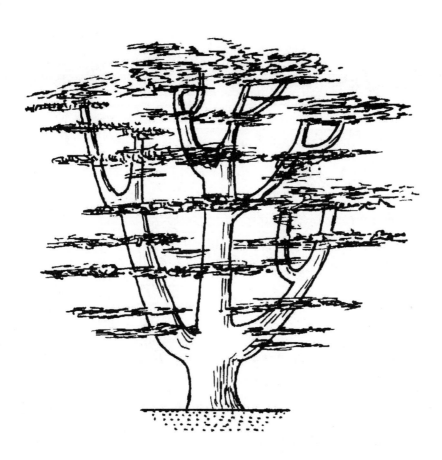

Growth Forms 1
生长型 1

Procumbent
平铺的

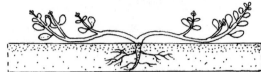

Decumbent
外倾的

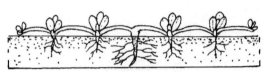

Repent
匍匐（生根）的

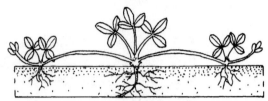

Stoloniferous
具匍匐茎的

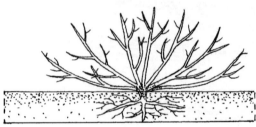

Ascending
斜升的

Clockwise climbing
shoot of *Lonicera* sp.
忍冬属植物顺时针
方向的攀缘枝

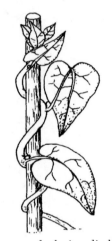

Counter–clockwise climbing
shoot of *Convolvulus* sp.
旋花属植物的逆时针方向
攀缘枝

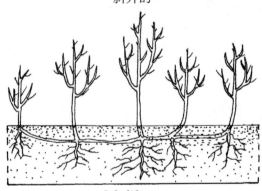

Soboliferous
具根出条的

Subshrub
亚灌木

Shrub
灌木

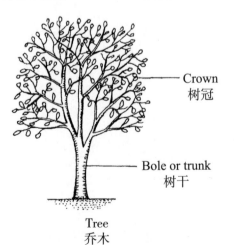

Crown
树冠

Bole or trunk
树干

Tree
乔木

Growth Forms 2
生长型 2

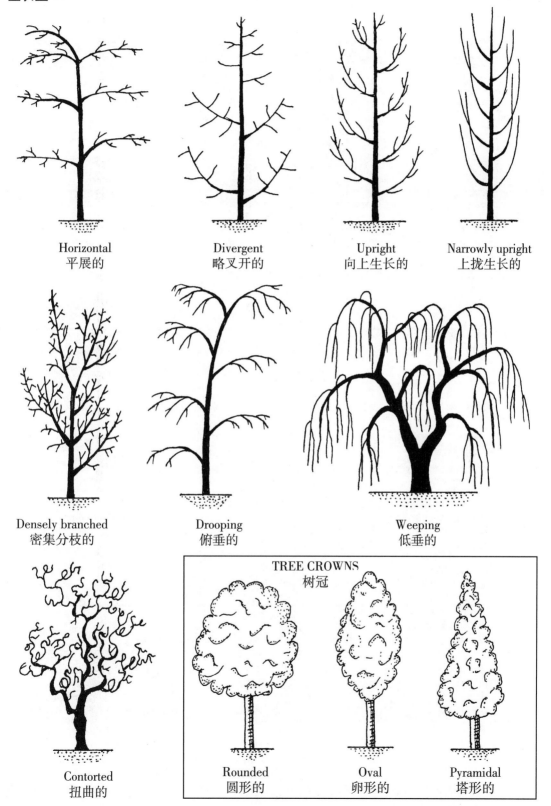

Horizontal
平展的

Divergent
略叉开的

Upright
向上生长的

Narrowly upright
上拢生长的

Densely branched
密集分枝的

Drooping
俯垂的

Weeping
低垂的

Contorted
扭曲的

TREE CROWNS
树冠

Rounded
圆形的

Oval
卵形的

Pyramidal
塔形的

Growth Forms 3
生长型 3

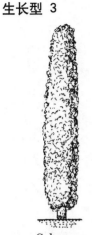

Columnar
(*Calocedrus decurrens*)
柱状的
（北美翠柏）

Fastigiate
(*Taxus baccata* 'Fastigiata')
帚状的
（欧洲红豆杉）

Pyramidal with oblique
branches
(*Picea abies*)
具斜枝的塔形的
（挪威云杉）

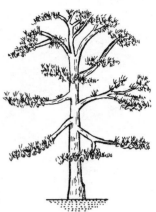

Flat–topped
(*Pinus sylvestris*)
平顶的
（长白松）

Drooping branches
(*Cedrus deodara*)
俯垂枝（雪松）

Oblique or ascending branches
(*Picea sitchensis*)
斜或斜升枝（北美云杉）

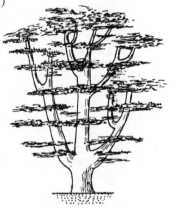

Horizontal branches
(*Cedrus libani*)
平展枝（黎巴嫩雪松）

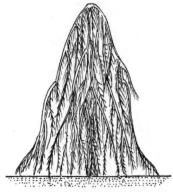

Pendulous branches
(*Cedrus atlantica* 'Glauca Pendula')
悬垂枝
（蓝叶垂枝北非雪松）

Conical
(*Chamaecyparis* lawsoniana
'Winston Churchill')
球果状的（美国扁柏"丘吉尔"）

Prostrate
(*Taxus baccata* 'Dovastoniana')
平卧的（欧洲红豆杉）

Growth Forms 4
生长型 4

Acaulescent (*Cirsium acaule*)
无茎的（无茎蓟）

Cauline leaf
茎生叶

Caulescent (*Stellaria* sp.)
具茎的（繁缕属植物）

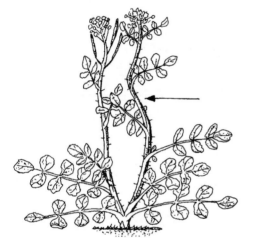

Flexuous (*Cardamine flexuosa*)
"之"字形的（弯曲碎米荠）

Caespitose (*Deschampsia cespitosa*)
丛生的（发草）

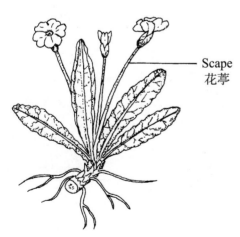

Scape
花葶

Scapose (*Primula vulgaris*)
具花葶的（欧洲报春）

Burs with epicormic shoots on *Tilia x europaea*
欧洲椴的具外生球茎枝条的刺球体

Life Forms (Raunkiaer's Classification)
生活型（瑙基耶尔分类）

The perennating parts of plants are shown in black
植物的多年生部分以黑体示出

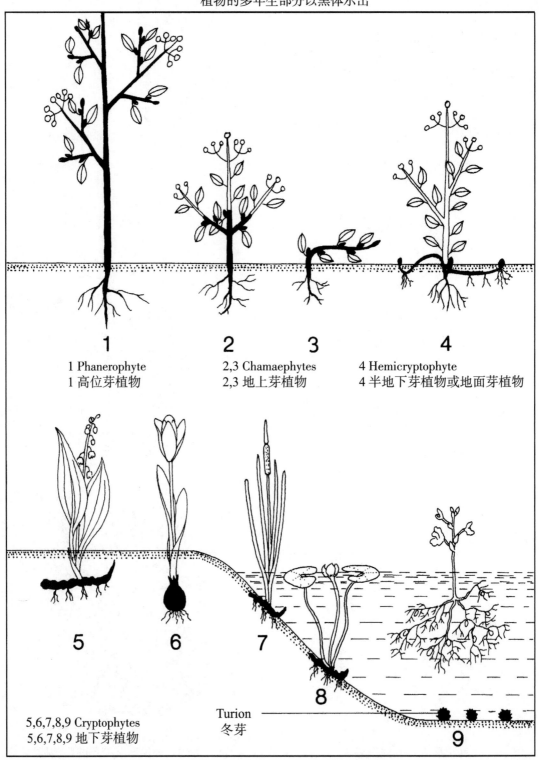

1 Phanerophyte
1 高位芽植物

2,3 Chamaephytes
2,3 地上芽植物

4 Hemicryptophyte
4 半地下芽植物或地面芽植物

5,6,7,8,9 Cryptophytes
5,6,7,8,9 地下芽植物

Turion
冬芽

4. General Features of Flowering Plants
4. 有花植物的一般特征

Parts of a Typical Monocotyledon
典型单子叶植物部分

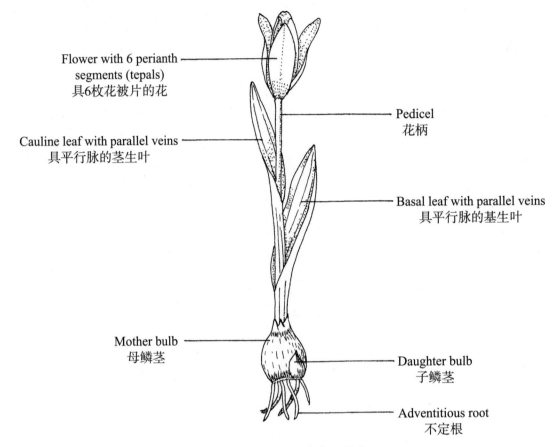

Flower with 6 perianth
segments (tepals)
具6枚花被片的花

Cauline leaf with parallel veins
具平行脉的茎生叶

Pedicel
花柄

Basal leaf with parallel veins
具平行脉的基生叶

Mother bulb
母鳞茎

Daughter bulb
子鳞茎

Adventitious root
不定根

Main parts of a monocotyledon: *Tulipa* sp.
单子叶植物的主要部分：郁金香属某植物

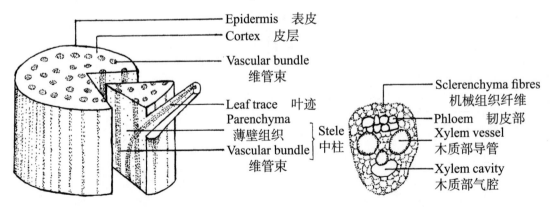

Epidermis　表皮
Cortex　皮层
Vascular bundle
维管束

Leaf trace　叶迹
Parenchyma
薄壁组织
Vascular bundle
维管束

Stele
中柱

Sclerenchyma fibres
机械组织纤维
Phloem　韧皮部
Xylem vessel
木质部导管
Xylem cavity
木质部气腔

Structure of a monocotyledon stem: *Zea mays*
单子叶植物茎的结构：玉米

Simplified diagram of a vascular bundle of *Zea
mays* (× 300)
玉米维管束简图（× 300）

Parts of a Typical Herbaceous Dicotyledon
典型草本双子叶植物部分

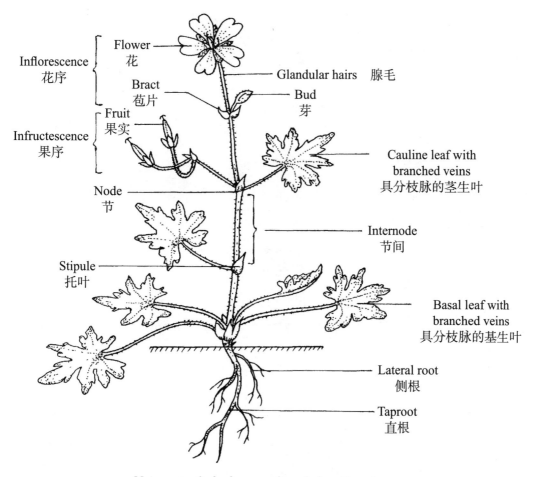

Main parts of a herbaceous dicotyledon: *Geranium* sp.
草本双子叶植物的主要部分：老鹳草属某植物

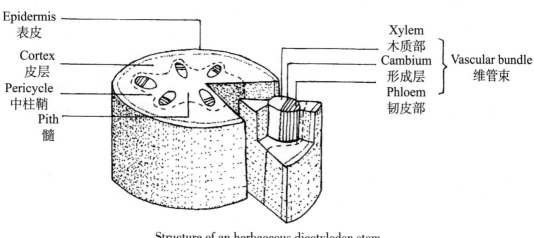

Structure of an herbaceous dicotyledon stem
草本双子叶植物茎的结构

Parts of a Winter Twig
冬枝部分

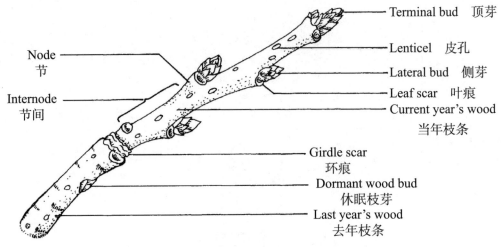

Terminal bud　顶芽
Lenticel　皮孔
Lateral bud　侧芽
Leaf scar　叶痕
Current year's wood
当年枝条
Node
节
Internode
节间
Girdle scar
环痕
Dormant wood bud
休眠枝芽
Last year's wood
去年枝条

Stem of a woody dicotyledon: *Prunus* sp.
木本双子叶植物的茎：李属某植物

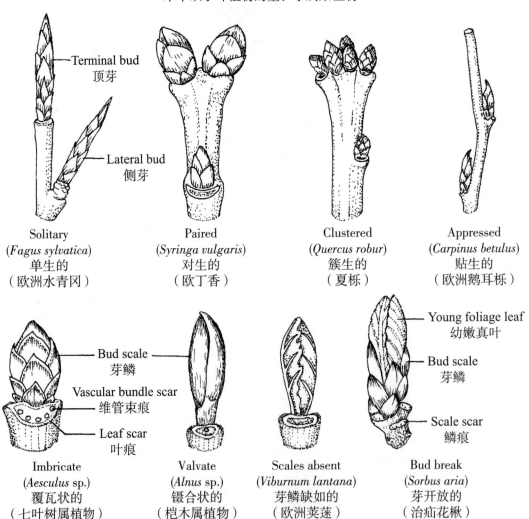

Terminal bud
顶芽
Lateral bud
侧芽

Solitary
(*Fagus sylvatica*)
单生的
（欧洲水青冈）

Paired
(*Syringa vulgaris*)
对生的
（欧丁香）

Clustered
(*Quercus robur*)
簇生的
（夏栎）

Appressed
(*Carpinus betulus*)
贴生的
（欧洲鹅耳枥）

Bud scale
芽鳞
Vascular bundle scar
维管束痕
Leaf scar
叶痕

Young foliage leaf
幼嫩真叶
Bud scale
芽鳞
Scale scar
鳞痕

Imbricate
(*Aesculus* sp.)
覆瓦状的
（七叶树属植物）

Valvate
(*Alnus* sp.)
镊合状的
（桤木属植物）

Scales absent
(*Viburnum lantana*)
芽鳞缺如的
（欧洲荚蒾）

Bud break
(*Sorbus aria*)
芽开放的
（治疝花楸）

Structure of a Woody Dicotyledon Stem
木本双子叶植物茎的结构

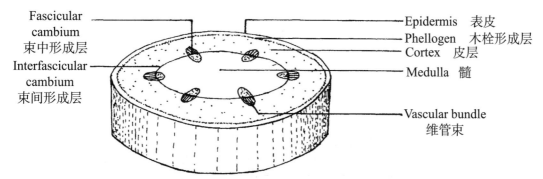

Fascicular cambium 束中形成层
Interfascicular cambium 束间形成层

Epidermis　表皮
Phellogen　木栓形成层
Cortex　皮层
Medulla　髓

Vascular bundle
维管束

T.S. of a dicotyledon stem at the start of secondary thickening
次生加厚开始时的双子叶植物茎的横切

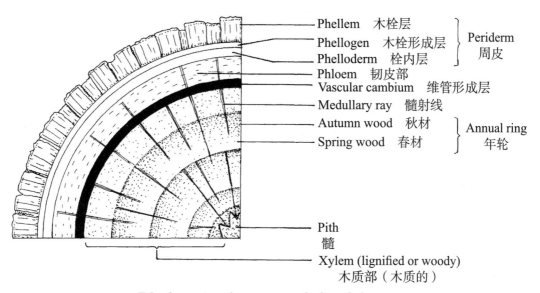

Phellem　木栓层
Phellogen　木栓形成层
Phelloderm　栓内层
⎫ Periderm
⎬ 周皮
⎭

Phloem　韧皮部
Vascular cambium　维管形成层
Medullary ray　髓射线
Autumn wood　秋材
Spring wood　春材
⎫ Annual ring
⎬ 年轮
⎭

Pith
髓

Xylem (lignified or woody)
木质部（木质的）

T.S. of a portion of a mature woody dicotyledon stem
成熟木本双子叶植物茎的部分横切

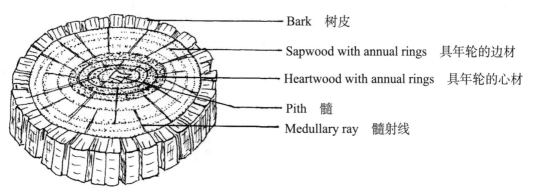

Bark　树皮

Sapwood with annual rings　具年轮的边材

Heartwood with annual rings　具年轮的心材

Pith　髓

Medullary ray　髓射线

Gross structure of a log of wood
一段木材的整体结构

5. Plant Features and Responses
5. 植物特征和响应

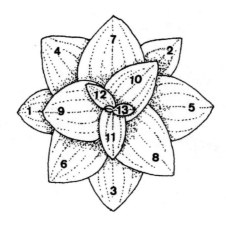

Stem Shapes, Leaf Fall, Apical Meristem, Phototropism
茎的形状、叶落、顶端分生组织、趋光性

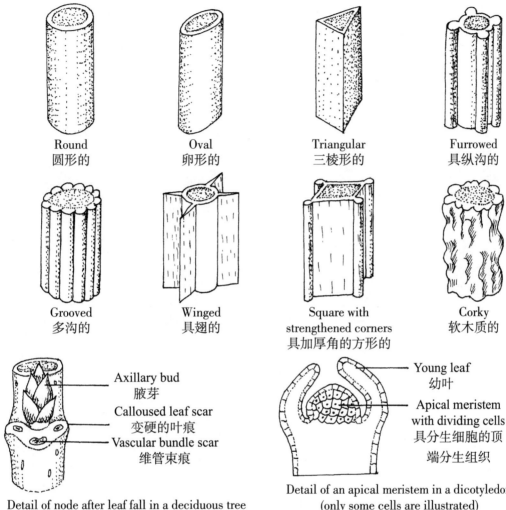

Round
圆形的

Oval
卵形的

Triangular
三棱形的

Furrowed
具纵沟的

Grooved
多沟的

Winged
具翅的

Square with
strengthened corners
具加厚角的方形的

Corky
软木质的

Axillary bud
腋芽

Calloused leaf scar
变硬的叶痕

Vascular bundle scar
维管束痕

Detail of node after leaf fall in a deciduous tree
落叶乔木叶落后节的详图

Young leaf
幼叶

Apical meristem
with dividing cells
具分生细胞的顶
端分生组织

Detail of an apical meristem in a dicotyledon
(only some cells are illustrated)
双子叶植物顶端分生组织详图
（仅绘出部分细胞）

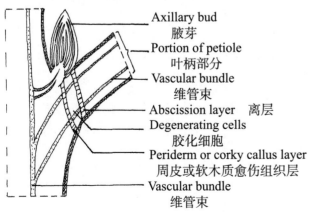

Axillary bud
腋芽

Portion of petiole
叶柄部分

Vascular bundle
维管束

Abscission layer　离层

Degenerating cells
胶化细胞

Periderm or corky callus layer
周皮或软木质愈伤组织层

Vascular bundle
维管束

Leaf fall in a woody dicotyledon
木本双子叶植物的叶落机制

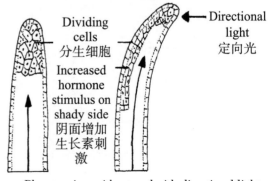

Dividing
cells
分生细胞

Increased
hormone
stimulus on
shady side
阴面增加
生长素刺
激

Directional
light
定向光

Phototropism without and with directional light
(arrows denote direction of growth)
无定向光和有定向光的趋光性
（箭头指示生长方向）

Parts Responses
植物响应

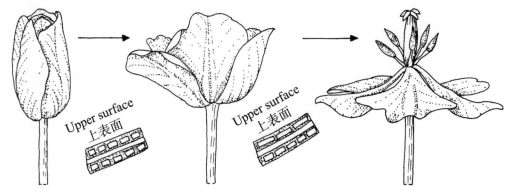

Thermonasty. Stages in flower opening in *Tulipa* sp. in cool to warm conditions
感热性：在自冷到热条件下郁金香属植物的开花期

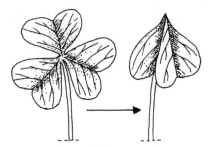

Nyctinasty in *Oxalis* sp.
酢浆草属植物的感夜性

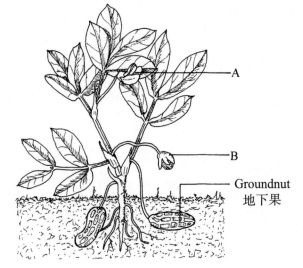

Geocarpy in *Arachis hypogaea*. A. Flower before pollination
花生的地下结果性 传粉前的花
B. After pollination, when the pedicel is negatively
heliotropic and positively geotropic
传粉后，花梗具有负的向光性和正的向地性

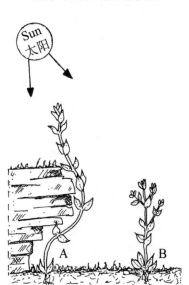

Etiolation. A. Plant growing in
shade under a rock edge
B. Same plant in full sunlight
黄化性：A. 生长在岩石边缘下
背阴处的植物
B. 在充足阳光下的同一植物

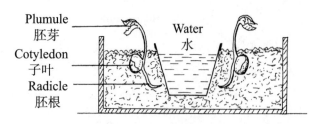

Hydrotropism. *Vicia faba* seeds germinating
around a water-filled porous pot
向水性：蚕豆种子在装满水的具孔的
花盆周围萌发

Phyllotaxy, Fibonacci Series, Leaf Mosaic, Heterophylly
叶序、斐波纳契数列、叶镶嵌、异形叶性

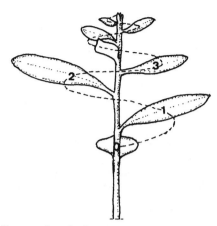

Phyllotaxy–the ideal spiral arrangement for leaves and branches (facial view)
叶序——理想的叶和枝螺旋状排列（侧面观）

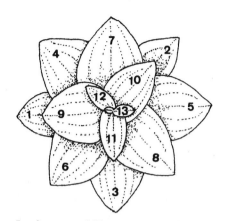

Leaf rosette of *Plantago major* showing phyllotaxy
大车前的莲座状叶，示叶序

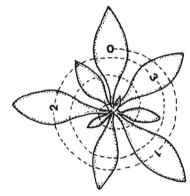

Phyllotaxy–the ideal spiral arrangement for leaves and branches (aerial view)
叶序——理想的叶和枝螺旋状排列（顶面观）

Leaf mosaic of *Acer pseudoplatanus*
欧亚槭的叶镶嵌

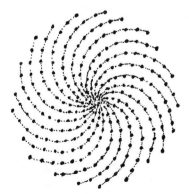

Fibonacci series or numbers, showing arrangement of disc florets in *Helianthus annuus* (each dot represents a floret)
斐波纳契数列或数字，示向日葵盘花的排列（每一点表示一朵小花）

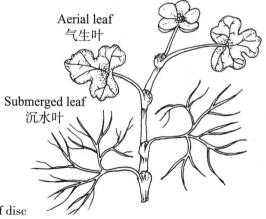

Aerial leaf
气生叶

Submerged leaf
沉水叶

Heterophylly in the aquatic plant *Ranunculus aquatilis*
水生植物水毛茛的异型叶性

6. Leaf-like Structures and Other Vegetative Features
6. 叶状结构和其他营养特征

Petioles, Stipules, Spines, Tendrils 1
叶柄、托叶、刺、卷须 1

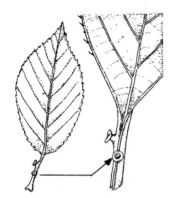

Petiole of *Prunus* sp. with glands alternate
李属植物具互生腺体的叶柄

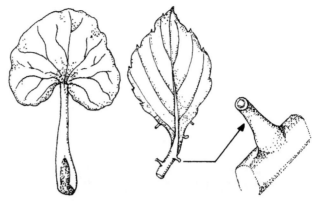

Sheathing petiole
(*Ranunculus ficaria*)
鞘状叶柄（榕叶毛莨）

Petiole with nectariferous glands
(*Impatiens walleriana*)
具蜜腺的叶柄（苏丹凤仙花）

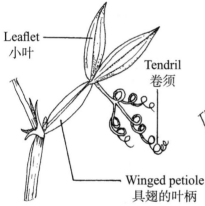

Leaflet
小叶
Tendril
卷须
Winged petiole
具翅的叶柄

Winged petiole (*Lathyrus odoratus*)
具翅的叶柄（香豌豆）

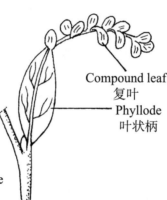

Compound leaf
复叶
Phyllode
叶状柄

Phyllode (*Acacia* sp.)
叶状柄（金合欢属植物）

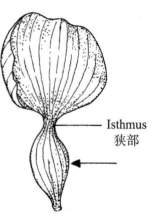

Isthmus
狭部

Swollen petiole
(*Eichhornia crassipes*)
膨大的叶柄（水葫芦）

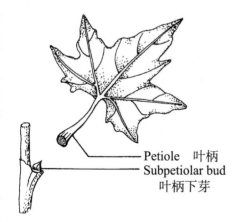

Petiole 叶柄
Subpetiolar bud
叶柄下芽

Protective petiole (*Platanus* x *hispanica*)
detached from the axillary bud
从腋芽脱离的保护性叶柄
（二球悬铃木）

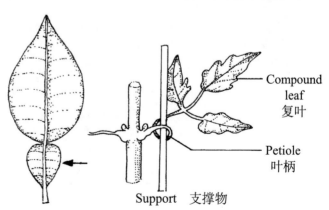

Compound
leaf
复叶
Petiole
叶柄
Support 支撑物

Winged petiole (*Citrus* sp.)
具翅的叶柄
（柑橘属植物）

Petiole of *Clematis* sp. adapted
for clinging to a support
适于攀缘到支撑物上的铁
线莲属植物的叶柄

Petioles, Stipules, Spines, Tendrils 2
叶柄、托叶、刺、卷须 2

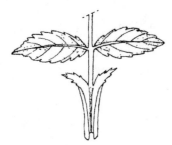

Adnate leafy stipules (*Rosa* sp.)
贴生的叶状托叶
（蔷薇属植物）

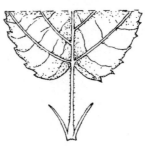

Filiform stipules
(*Corylus avellana*)
线状托叶（欧洲榛）

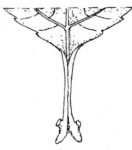

Petiole of *Prunus* sp.
with glands opposite
具对生腺体的李属植物的叶柄

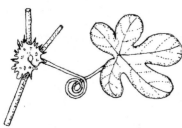

Fringed stipule
(*Tropaeolum ciliatum*)
流苏状托叶（缘毛旱金莲）

Connate stipules
(*Humulus lupulus*)
合生托叶（忽布）

Pinnately divided stipules
(*Viola tricolor*)
羽状裂的托叶（三色堇）

Axillary stipules
(*Boehmeria nivea*)
腋生的托叶（苎麻）

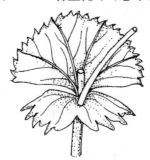

Antidromous stipules
(*Alchemilla mollis*)
边缘相邻的托叶（羽衣草）

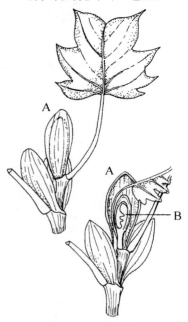

Sheathing stipules
(*Maranta* sp.)
鞘状托叶（竹芋属植物）

Connate stipules
(*Onobrychis viciifolia*)
合生托叶（驴食草）

Protective stipules
Bud (B) protected by stipules (A)
(*Liriodendron tulipifera*)
保护性托叶
被托叶（A）保护的芽（B）
（北美鹅掌楸）

Petioles, Stipules, Spines, Tendrils 3
叶柄、托叶、刺、卷须 3

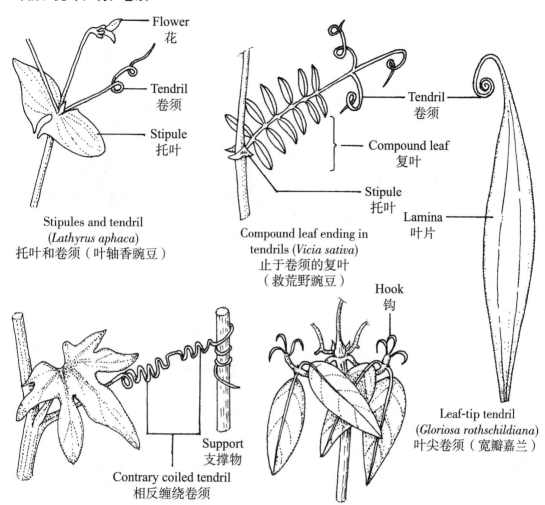

Stipules and tendril
(*Lathyrus aphaca*)
托叶和卷须（叶轴香豌豆）

Compound leaf ending in
tendrils (*Vicia sativa*)
止于卷须的复叶
（救荒野豌豆）

Coiling tendril (*Bryonia dioica*)
缠绕卷须（泻根）

Hook-like tendril (*Bignonia unguis–cati*)
钩状卷须（猫爪藤）

Leaf-tip tendril
(*Gloriosa rothschildiana*)
叶尖卷须（宽瓣嘉兰）

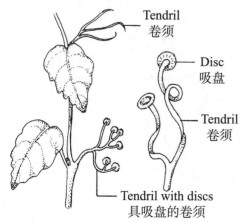

Tendril with and without sucker discs (*Parthenocissus* sp.)
具吸盘和不具吸盘的卷须（地锦属植物）

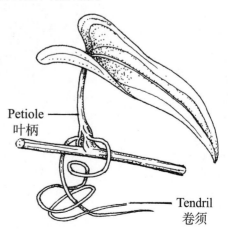

Stipular tendril (*Smilax aspera*)
托叶卷须（穗菝葜）

Petioles, Stipules, Spines, Tendrils 4
叶柄、托叶、刺、卷须 4

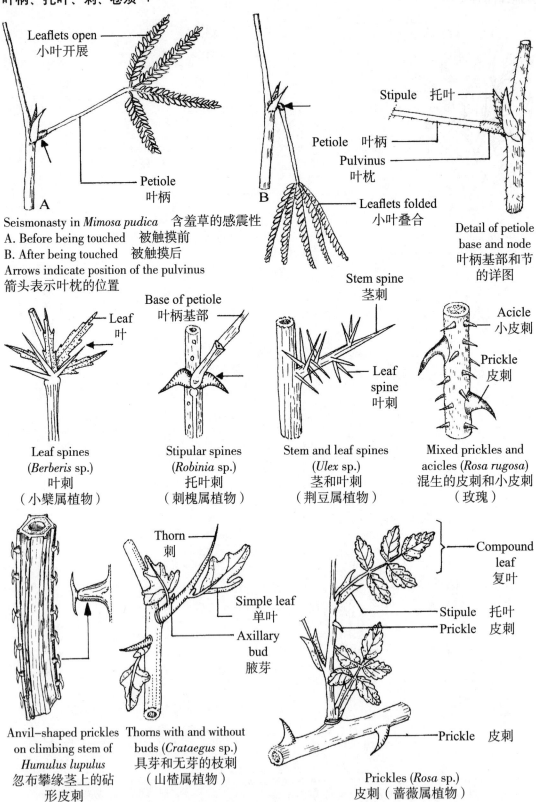

Leaflets open
小叶开展

Petiole
叶柄

A

Seismonasty in *Mimosa pudica*　含羞草的感震性
A. Before being touched　被触摸前
B. After being touched　被触摸后
Arrows indicate position of the pulvinus
箭头表示叶枕的位置

B

Leaflets folded
小叶叠合

Stipule　托叶
Petiole　叶柄
Pulvinus
叶枕

Detail of petiole
base and node
叶柄基部和节
的详图

Leaf
叶

Leaf spines
(*Berberis* sp.)
叶刺
（小檗属植物）

Base of petiole
叶柄基部

Stipular spines
(*Robinia* sp.)
托叶刺
（刺槐属植物）

Stem spine
茎刺

Leaf
spine
叶刺

Stem and leaf spines
(*Ulex* sp.)
茎和叶刺
（荆豆属植物）

Acicle
小皮刺

Prickle
皮刺

Mixed prickles and
acicles (*Rosa rugosa*)
混生的皮刺和小皮刺
（玫瑰）

Anvil–shaped prickles
on climbing stem of
Humulus lupulus
忽布攀缘茎上的砧
形皮刺

Thorn
刺

Simple leaf
单叶

Axillary
bud
腋芽

Thorns with and without
buds (*Crataegus* sp.)
具芽和无芽的枝刺
（山楂属植物）

Compound
leaf
复叶

Stipule　托叶
Prickle　皮刺

Prickle　皮刺

Prickles (*Rosa* sp.)
皮刺（蔷薇属植物）

Other Leaf Features
其他.叶的特征

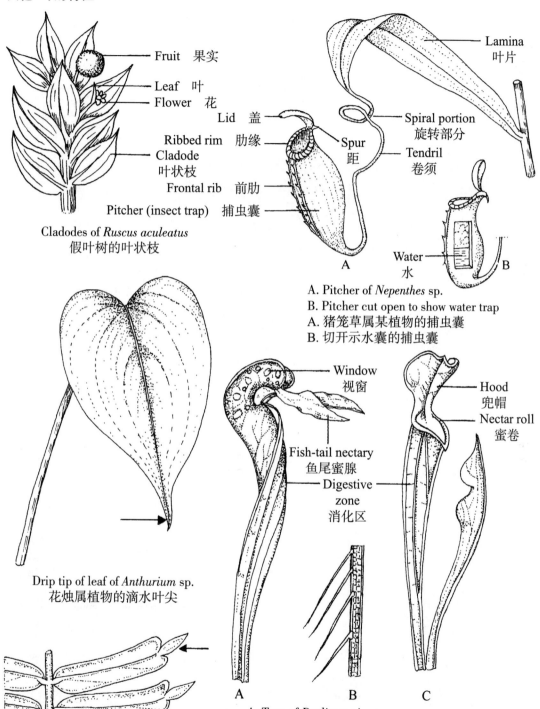

Fruit　果实

Leaf　叶

Flower　花

Lid　盖

Ribbed rim　肋缘

Cladode
叶状枝

Frontal rib　前肋

Pitcher (insect trap)　捕虫囊

Spur
距

Lamina
叶片

Spiral portion
旋转部分

Tendril
卷须

Water
水

Cladodes of *Ruscus aculeatus*
假叶树的叶状枝

A. Pitcher of *Nepenthes* sp.
B. Pitcher cut open to show water trap
A. 猪笼草属某植物的捕虫囊
B. 切开示水囊的捕虫囊

Drip tip of leaf of *Anthurium* sp.
花烛属植物的滴水叶尖

Window
视窗

Fish-tail nectary
鱼尾蜜腺

Digestive
zone
消化区

Hood
兜帽

Nectar roll
蜜卷

Beltian bodies terminating leaflets
of *Acacia cornigera*
牛角相思树小叶顶端的贝尔特体

A. Trap of *Darlingtonia* sp.
B. Downward pointing hairs in trap of *Darlingtonia* sp.
C. Trap of *Sarracenia* sp.
A. 眼镜蛇草属植物的捕虫囊
B. 眼镜蛇草属植物捕虫囊的倒向毛
C. 瓶子草属植物的捕虫囊

Domatia and Galls
虫穴和虫瘿

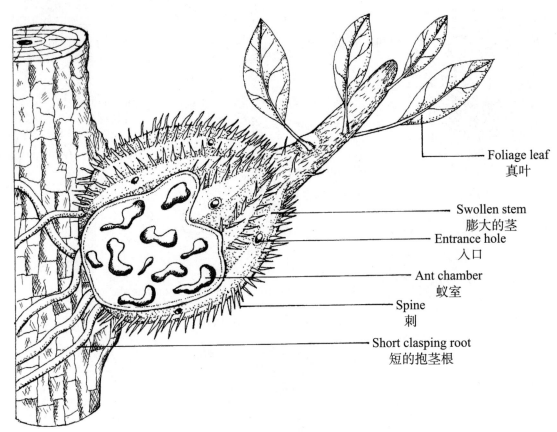

Myrmecodia sp., an epiphytic ant–plant. Portion cut open to show chambers
蚁巢木属植物，一种附生有蚂蚁的植物，切开部分示蚁室

Foliage leaf
真叶

Swollen stem
膨大的茎

Entrance hole
入口

Ant chamber
蚁室

Spine
刺

Short clasping root
短的抱茎根

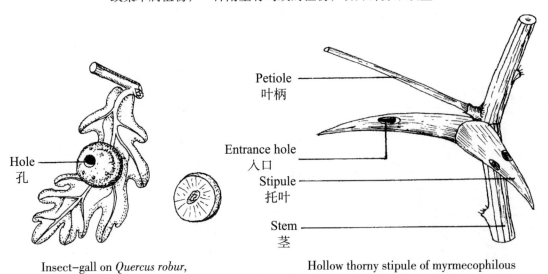

Hole
孔

Petiole
叶柄

Entrance hole
入口

Stipule
托叶

Stem
茎

Insect–gall on *Quercus robur*,
whole and in section
夏栎上的虫瘿，整体和部分

Hollow thorny stipule of myrmecophilous
plant, *Acacia nicoyensis*
蚁喜植物，角刺金合欢的空刺状托叶

Abnormal Forms
非正常型

Hose-in-hose *Primula*
报春花属植物的管中管花

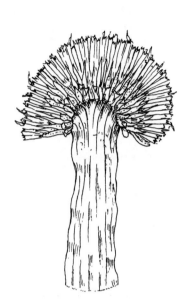

Fasciation in an inflorescence of
Taraxacum sp.
蒲公英属植物花序的扁化

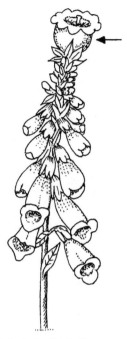

Peloria in *Digitalis purpurea*
毛地黄的非正常整齐花

Fasciation in a *Forsythia* stem
连翘属植物茎的扁化

Witches' broom on *Betula pendula*
垂枝桦的丛枝病

Bracts
苞片

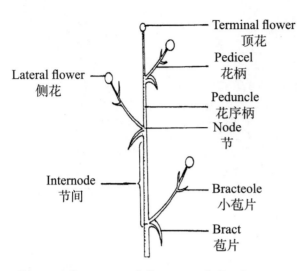

Diagram of a racemose inflorescence indicating
the position of bracts and bracteoles
总状花序的图解示苞片和小苞片的位置

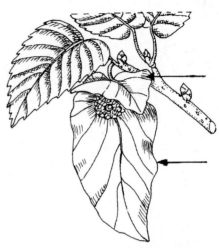

Paired bracts subtending the
inflorescence of *Davidia involucrata*
珙桐包着花序的成对苞片

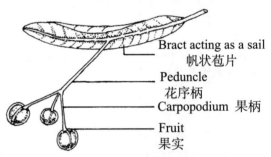

Adnate bract of *Tilia platyphyllos*
阔叶椴的贴生苞片

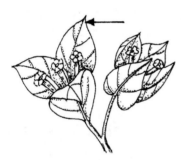

Petaloid bracts subtending the flowers of
Bougainvillea glabra
叶子花的包着花的花瓣状苞片

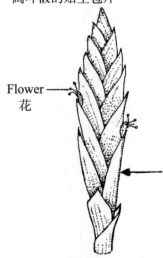

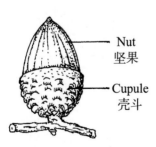

Fused cupular bracts on the
acorn of *Quercus robur*
夏栎坚果的愈合的杯状苞片

Involucral bracts of
Cynara scolymus
菜蓟的总苞

Imbricate bracts on the spicate
inflorescence of *Vriesia splendens*
美丽花叶兰穗状花序上的覆
瓦状苞片

7. Leaves
7. 叶

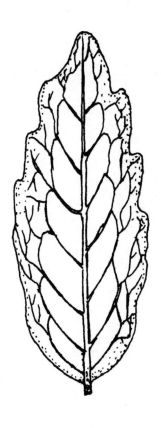

Leaf Arrangements
叶的排列

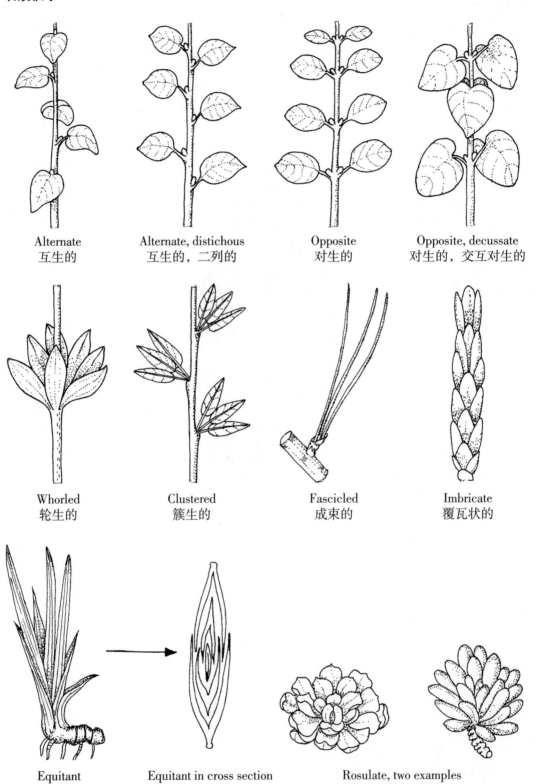

Alternate
互生的

Alternate, distichous
互生的，二列的

Opposite
对生的

Opposite, decussate
对生的，交互对生的

Whorled
轮生的

Clustered
簇生的

Fascicled
成束的

Imbricate
覆瓦状的

Equitant
套折的

Equitant in cross section
套折的（横切）

Rosulate, two examples
莲座状的，两例

Points of Attachment
附着点

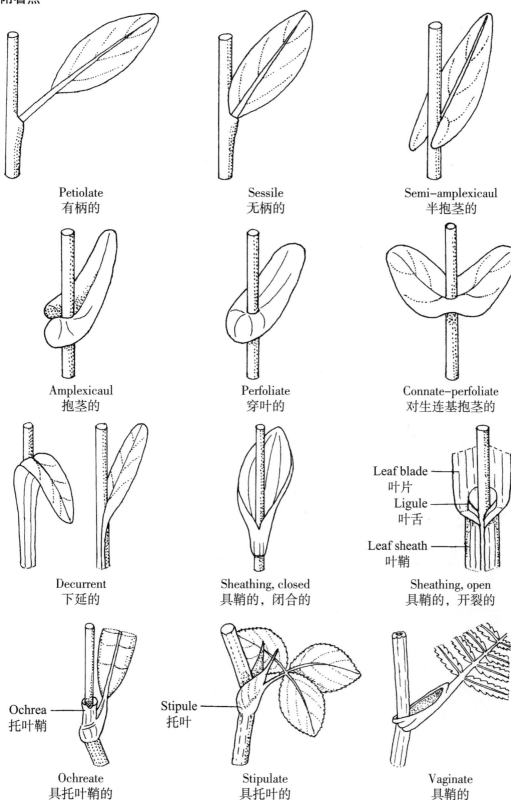

Petiolate
有柄的

Sessile
无柄的

Semi-amplexicaul
半抱茎的

Amplexicaul
抱茎的

Perfoliate
穿叶的

Connate-perfoliate
对生连基抱茎的

Leaf blade
叶片

Ligule
叶舌

Leaf sheath
叶鞘

Decurrent
下延的

Sheathing, closed
具鞘的，闭合的

Sheathing, open
具鞘的，开裂的

Ochrea
托叶鞘

Stipule
托叶

Ochreate
具托叶鞘的

Stipulate
具托叶的

Vaginate
具鞘的

A Typical Leaf and its Attachment
典型叶和它的附着物

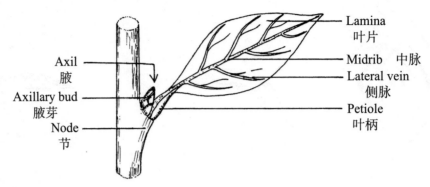

Axil
腋

Axillary bud
腋芽

Node
节

Lamina
叶片

Midrib　中脉

Lateral vein
侧脉

Petiole
叶柄

Leaf Folding including Ptyxis
包括卷叠式的叶折叠

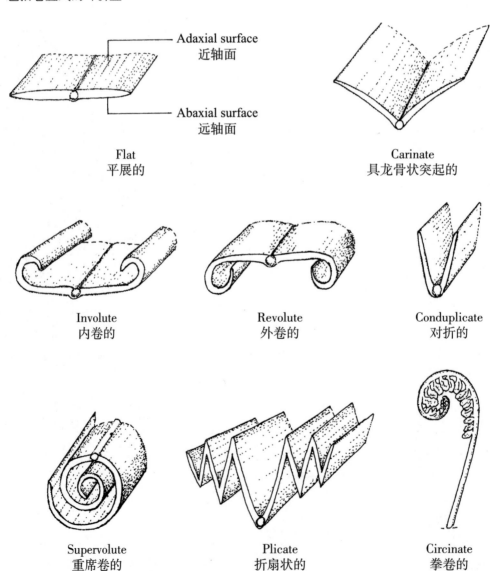

Adaxial surface
近轴面

Abaxial surface
远轴面

Flat
平展的

Carinate
具龙骨状突起的

Involute
内卷的

Revolute
外卷的

Conduplicate
对折的

Supervolute
重席卷的

Plicate
折扇状的

Circinate
拳卷的

Vernation
幼叶卷迭式

Imbricate
覆瓦状的

Equitant
套折的

Convolute
旋转的

Venation
脉序

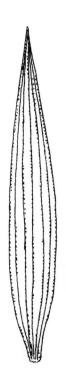

Parallel
平行的

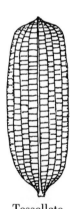

Tessellate
棋盘格状的

Anastomosing
网结的

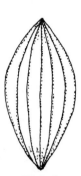

Bowed
弧形的

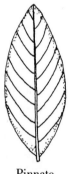

Pinnate
羽状的

Reticulate
网状的

Palmate, two forms
掌状的，两种形式

Simple Leaves 1
单叶 1

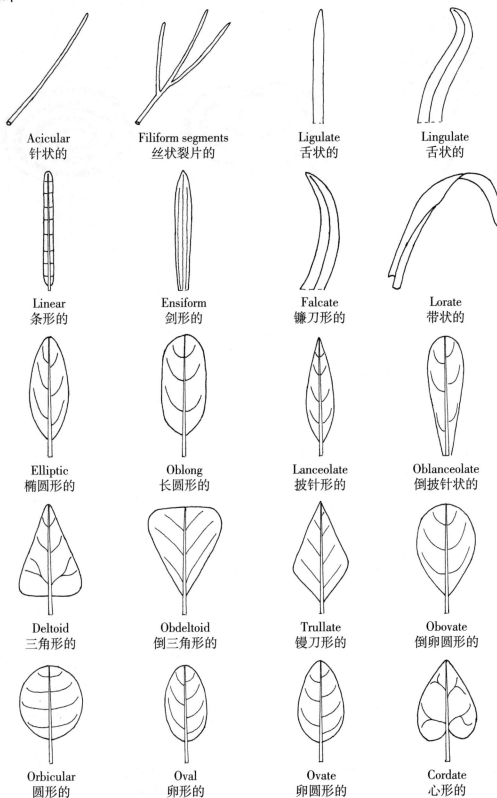

Acicular 针状的	Filiform segments 丝状裂片的	Ligulate 舌状的	Lingulate 舌状的
Linear 条形的	Ensiform 剑形的	Falcate 镰刀形的	Lorate 带状的
Elliptic 椭圆形的	Oblong 长圆形的	Lanceolate 披针形的	Oblanceolate 倒披针状的
Deltoid 三角形的	Obdeltoid 倒三角形的	Trullate 镘刀形的	Obovate 倒卵圆形的
Orbicular 圆形的	Oval 卵形的	Ovate 卵圆形的	Cordate 心形的

Simple Leaves 2
单叶 2

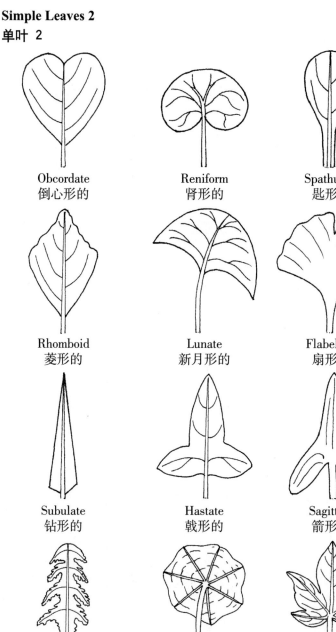

Obcordate
倒心形的

Reniform
肾形的

Spathulate
匙形的

Pandurate
提琴形的

Rhomboid
菱形的

Lunate
新月形的

Flabellate
扇形的

Cuneate
楔形的

Subulate
钻形的

Hastate
戟形的

Sagittate
箭形的

Lyrate
大头羽裂的

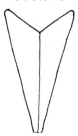

Runcinate
倒向羽裂的

Peltate
盾状的

Palmatifid
掌状裂的

Palmatisect
掌状全裂的

Pinnatifid
羽状半裂的

Pinnatisect
羽状全裂的

Pinnatipartie
羽状深裂的

Fan-shaped
扇形的

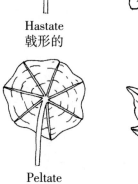

Compound Leaves
复叶

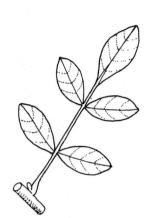

Imparipinnate (odd–pinnate)
奇数羽状的

Paripinnate (even–pinnate)
偶数羽状的

Bipinnate
二回羽状的

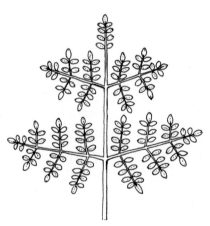

Tripinnate
三回羽状的

Biternate
二回三出的

Trifoliolate
具三小叶的

Pedate
鸟足状的

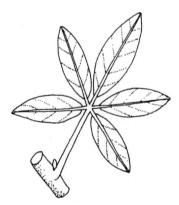

Palmate
掌状的

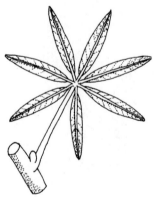

Digitate
指状的

Leaf Margins
叶缘

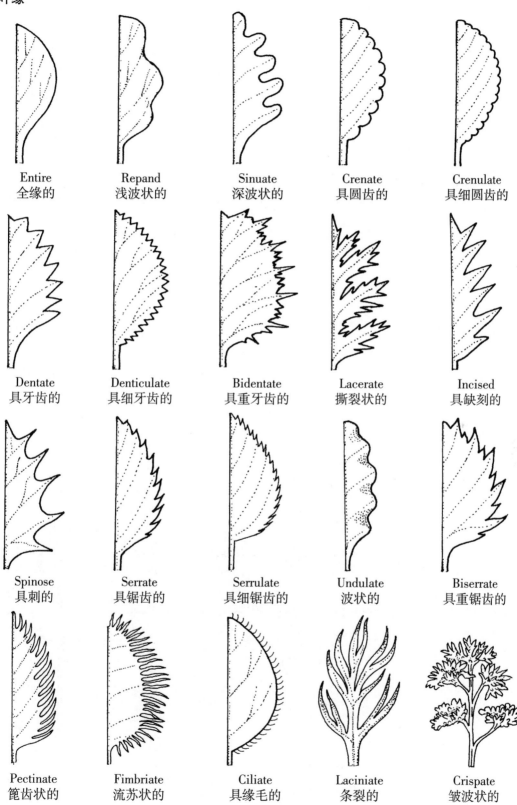

Entire
全缘的

Repand
浅波状的

Sinuate
深波状的

Crenate
具圆齿的

Crenulate
具细圆齿的

Dentate
具牙齿的

Denticulate
具细牙齿的

Bidentate
具重牙齿的

Lacerate
撕裂状的

Incised
具缺刻的

Spinose
具刺的

Serrate
具锯齿的

Serrulate
具细锯齿的

Undulate
波状的

Biserrate
具重锯齿的

Pectinate
篦齿状的

Fimbriate
流苏状的

Ciliate
具缘毛的

Laciniate
条裂的

Crispate
皱波状的

Leaf Bases
叶基

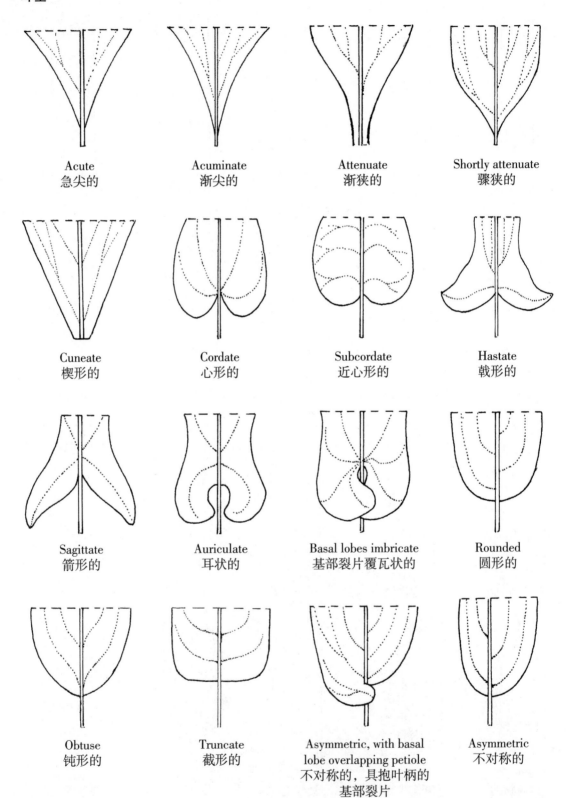

Acute
急尖的

Acuminate
渐尖的

Attenuate
渐狭的

Shortly attenuate
骤狭的

Cuneate
楔形的

Cordate
心形的

Subcordate
近心形的

Hastate
戟形的

Sagittate
箭形的

Auriculate
耳状的

Basal lobes imbricate
基部裂片覆瓦状的

Rounded
圆形的

Obtuse
钝形的

Truncate
截形的

Asymmetric, with basal
lobe overlapping petiole
不对称的，具抱叶柄的
基部裂片

Asymmetric
不对称的

Leaf Apices
叶尖

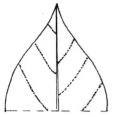
Acuminate
渐尖的

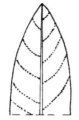
Acute
急尖的

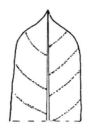
Abruptly acute
突尖的

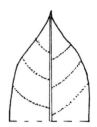
Apiculate
细尖的

Caudate
尾尖的

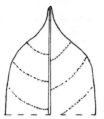
Cuspidate
骤尖的

Aristate
具芒的

Mucronate
短尖的

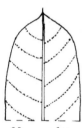
Mucronulate
小短尖的

Pungent
锐尖的

Hooked–truncate
截形具钩的

Cirrhose
卷须状的

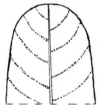
Rounded
圆形的

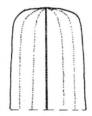
Truncate
截形的

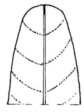
Obtuse
钝形的

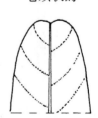
Retuse
微凹的

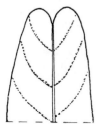
Emarginate
微缺的

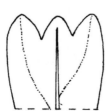
Tridentate
三牙齿的

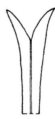
Cleft
半裂的

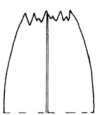
Praemorse
啮蚀状的

Leaf Structure 1
叶的结构 1

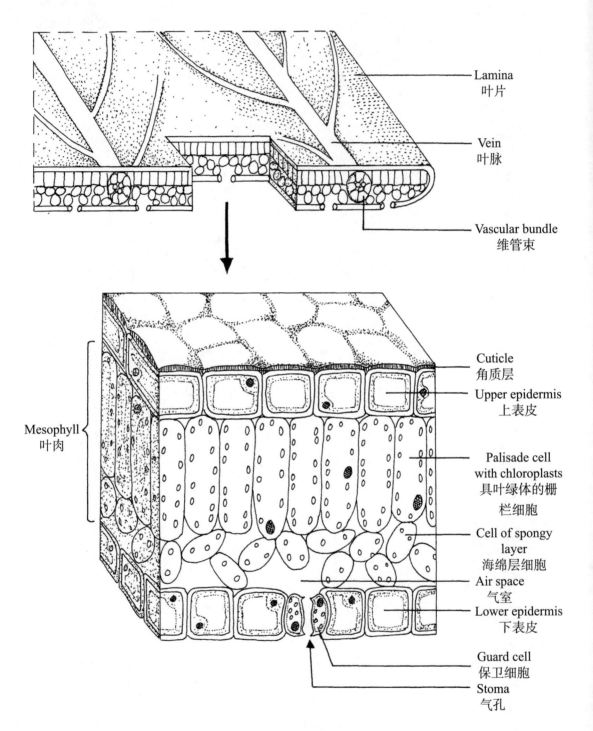

Lamina
叶片

Vein
叶脉

Vascular bundle
维管束

Mesophyll
叶肉

Cuticle
角质层

Upper epidermis
上表皮

Palisade cell
with chloroplasts
具叶绿体的栅
栏细胞

Cell of spongy
layer
海绵层细胞

Air space
气室

Lower epidermis
下表皮

Guard cell
保卫细胞

Stoma
气孔

Internal structure of a leaf
叶的内部结构

Leaf Structure 2
叶的结构 2

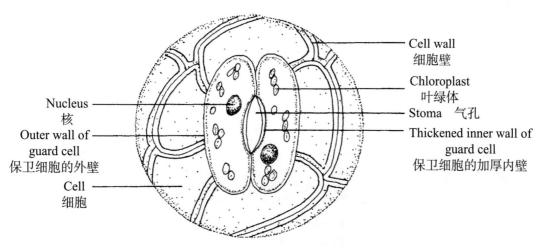

Nucleus 核
Outer wall of guard cell 保卫细胞的外壁
Cell 细胞

Cell wall 细胞壁
Chloroplast 叶绿体
Stoma 气孔
Thickened inner wall of guard cell 保卫细胞的加厚内壁

Lower epidermis of a leaf showing details of a typical stoma and guard cells
叶下表皮，示典型气孔和保卫细胞的详图

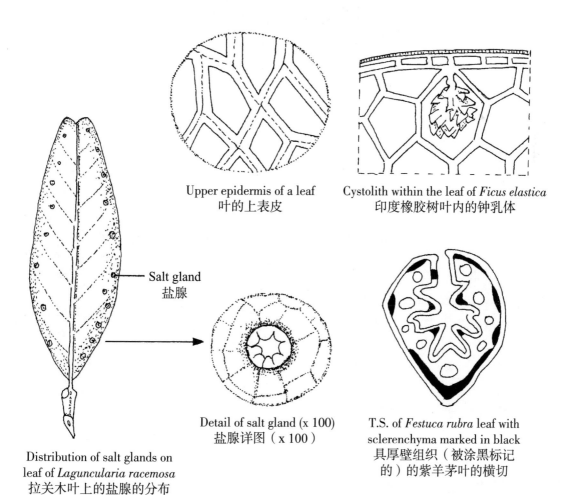

Upper epidermis of a leaf
叶的上表皮

Cystolith within the leaf of *Ficus elastica*
印度橡胶树叶内的钟乳体

Salt gland
盐腺

Detail of salt gland (x 100)
盐腺详图（x 100）

T.S. of *Festuca rubra* leaf with sclerenchyma marked in black
具厚壁组织（被涂黑标记的）的紫羊茅叶的横切

Distribution of salt glands on leaf of *Laguncularia racemosa*
拉关木叶上的盐腺的分布

8. Hairs and Scales
8. 毛和鳞片

Hairs and Scales 1
毛和鳞片 1

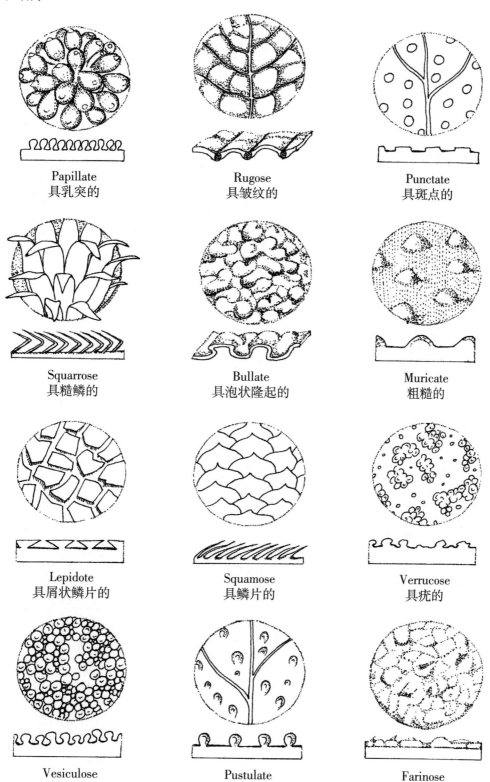

Papillate
具乳突的

Rugose
具皱纹的

Punctate
具斑点的

Squarrose
具糙鳞的

Bullate
具泡状隆起的

Muricate
粗糙的

Lepidote
具屑状鳞片的

Squamose
具鳞片的

Verrucose
具疣的

Vesiculose
泡状的

Pustulate
泡状突起的

Farinose
被粉的

Hairs and Scales 2
毛和鳞片 2

Scabrous
粗糙的

Setose
具刚毛的

Strigose
具糙伏毛的

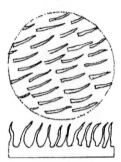

Hirsute
具长硬毛的

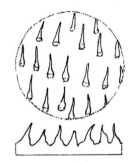

Hispid
具糙硬毛的

Paleaceous
糠秕状的

Pilose
具柔毛的

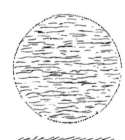

Sericeous
被绢毛的

Puberulous
被微柔毛的

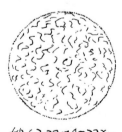

Tomentose
具绒毛的

Velutinous
被短绒毛的

Lanate
具绵毛的

Hairs and Scales 3
毛和鳞片 3

Villous
具长柔毛的

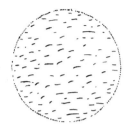

Canescent
被灰白毛的

Hirsutullous
具短硬毛的

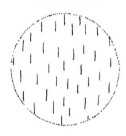

Hirtellous
具微硬毛的

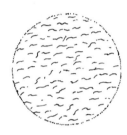

Pubescent
被短柔毛的

Stellate
星状的

Floccose
具丛卷毛的

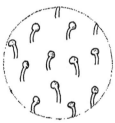

Glandular hairs
腺毛

Bifid hairs, 2 types
二裂毛，2种类型

Peltate
盾状的

Stellate hairs, 2 types
星状毛，2种类型

Hairs and Scales 4
毛和鳞片 4

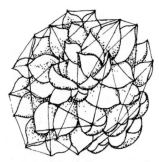

Arachnoid hairs (*Sempervivum arachnoideum*)
蛛丝状毛（蛛丝卷绢）

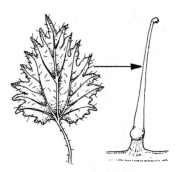

Stinging hairs (*Urtica dioica*)
螫毛（异株荨麻）

Simple (unicellular) hairs
(*Capsella bursa-pastoris*)
单（单细胞）毛（荠菜）

Dendritic hairs (*Mimosa* sp.)
枝状毛（含羞草属植物）

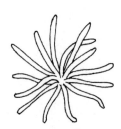

Hairs tuft (*Helianthemum* sp.)
毛簇（半月花属植物）

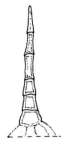

Multicellular hairs (*Olearia* sp.)
多细胞毛（榄叶菊属某植物）

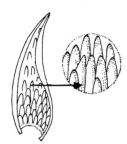

Colleter (*Cepallanthus* sp.)
黏液毛（凤箱树属植物）

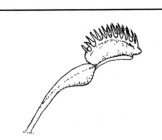

Whole leaf when closed
闭合时的整个叶片

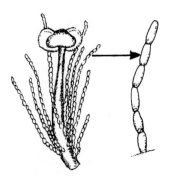

Moniliform hairs (*Tradescantia* sp.)
念珠状的毛（紫露草属植物）

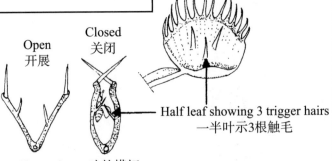

Open
开展

Closed
关闭

Leaf in section 叶的横切

Half leaf showing 3 trigger hairs
一半叶示3根触毛

Mechanism of trigger hairs (*Dionaea muscipula*)
触毛的机制（捕蝇草）

9. Floral Features
9. 花的特征

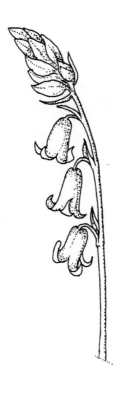

Inflorescences 1
花序 1

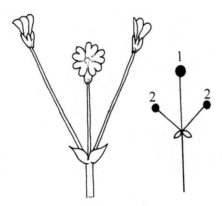

Simple cyme (*Cerastium* sp.)
简单聚伞花序（卷耳属植物）

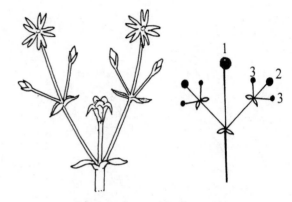

Compound cyme (*Stellaria* sp.)
复合聚伞花序（繁缕属植物）

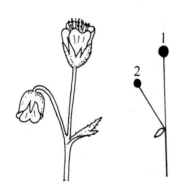

Simple cyme (*Geum* sp.)
简单聚伞花序（路边青属植物）

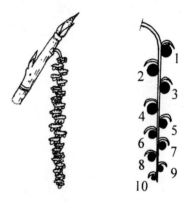

Catkin (*Populus* sp.)
柔荑花序（杨属植物）

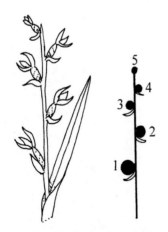

Spike (*Cephalanthera* sp.)
穗状花序（头蕊兰属植物）

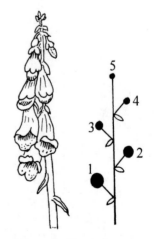

Raceme (*Digitalis* sp.)
总状花序（毛地黄属植物）

Inflorescences 2
花序 2

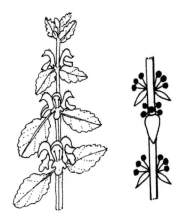

Verticillaster (*Lamium* sp.)
轮伞花序（野芝麻属植物）

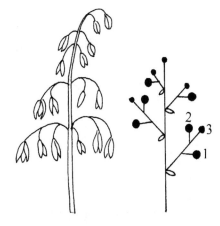

Panicle (*Avena* sp.)
圆锥花序（燕麦属植物）

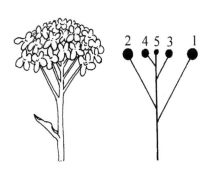

Corymb (*Iberis* sp.)
伞房花序（屈曲花属植物）

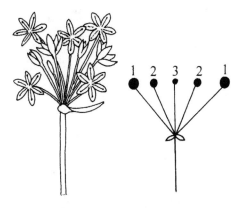

Simple umbel (*Allium* sp.)
简单伞形花序（葱属植物）

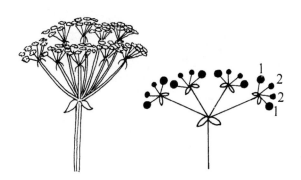

Compound umbel (*Heracleum* sp.)
复伞形花序（独活属植物）

Capitulum (*Doronicum* sp.)
头状花序（多榔菊属植物）

Inflorescences 3
花序 3

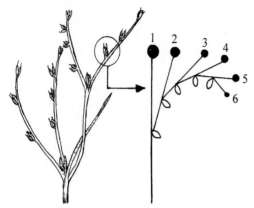

Drepanium (*Juncus bufonis*)
(also arrangement of drepania on plant)
镰状聚伞花序（小灯心草）
（在植株上也呈镰状聚伞花序排列）

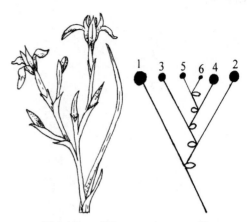

Rhipidium (*Moraea macgregorii*)
扇形聚伞花序（马氏肖鸢尾）

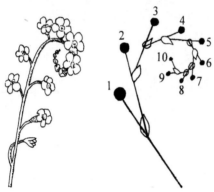

Cincinnus (*Myosotis scorpioides*)
蝎尾状聚伞花序（沼泽勿忘草）

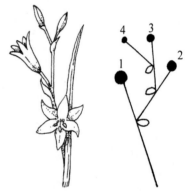

Bostryx (*Hemerocallis minor*)
螺状聚伞花序（小黄花菜）

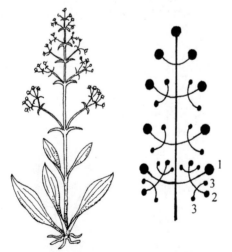

Thyrse (*Valeriana saxatilis*)
聚伞圆锥花序（石生缬草）

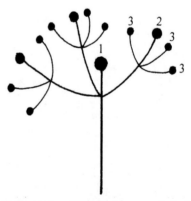

Pleiochasium (*Euphorbia cyparissias*)
多歧聚伞花序（欧洲柏大戟）

Inflorescences 4
花序 4

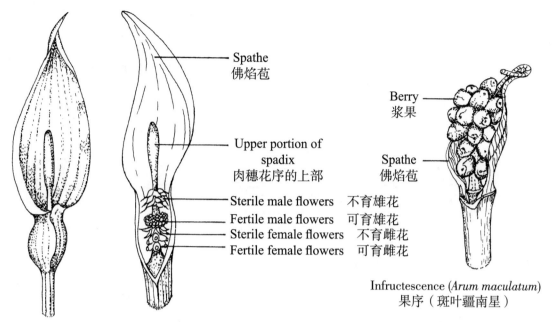

Spathe
佛焰苞

Upper portion of
spadix
肉穗花序的上部

Sterile male flowers　不育雄花
Fertile male flowers　可育雄花
Sterile female flowers　不育雌花
Fertile female flowers　可育雌花

Berry
浆果

Spathe
佛焰苞

Infructescence (*Arum maculatum*)
果序（斑叶疆南星）

Inflorescence (*Arum maculatum*)
花序（斑叶疆南星）

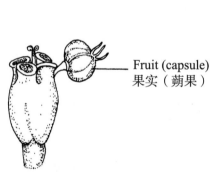

Fruit (capsule)
果实（蒴果）

Cyathium (*Euphorbia helioscopia*)
杯状聚伞花序（泽漆）

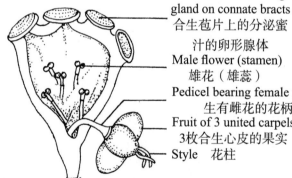

Oval nectar-secreting
gland on connate bracts
合生苞片上的分泌蜜
汁的卵形腺体
Male flower (stamen)
雄花（雄蕊）
Pedicel bearing female flower
生有雌花的花柄
Fruit of 3 united carpels
3枚合生心皮的果实
Style　花柱

Cyathium opened up (*Euphorbia helioscopia*)
展开的杯状聚伞花序（泽漆）

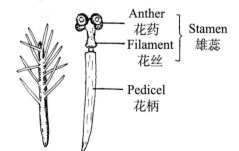

Anther
花药
Filament
花丝
Stamen
雄蕊

Pedicel
花柄

Male flower with its feathered bract
(*Euphorbia helioscopia*)
生有羽状苞片的雄花（泽漆）

Horned nectar-secretomg gland
(*Euphorbia cyparissias*)
角状分泌蜜汁的腺体（欧洲柏大戟）

Other Inflorescence and Flower Features 1
其他花序和花的特征 1

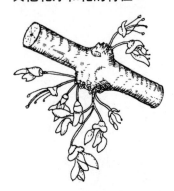

Cauliflory, showing flowers on
main stem (*Cercis siliquastrum*)
茎花现象，示主茎上的花
（南欧紫荆）

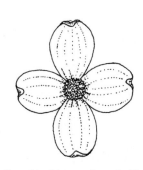

Pseudanthium subtended by
bracts (*Cornus florida*)
被苞片包着的假单花花序
（北美山茱萸）

Cluster of male flowers (*Fagus sylvatica*)
雄花簇（欧洲水青冈）

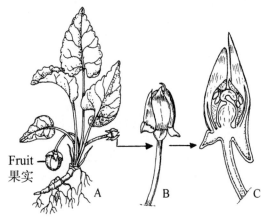

Fruit
果实

A. Whole plant of *Viola* sp. B. Cleistogamous flower
C. L.S. of flower (enlarged)
A. 堇菜属某植物的整株 B. 闭花受精的花
C. 花的纵切（放大的）

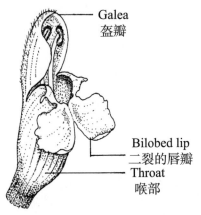

Galea
盔瓣

Bilobed lip
二裂的唇瓣
Throat
喉部

Upper portion of corolla (*Lamium purpureum*)
花冠的上部（紫花野芝麻）

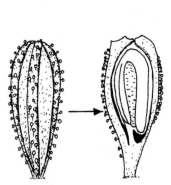

Anthocarp whole and in section
(*Boerhavia diffusa*)
掺花果的整体和纵切
（黄细心）

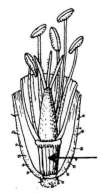

Anthophore (× 6)
(*Silene gallica*)
花冠柄（× 6）
（蝇子草）

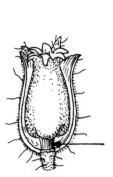

Carpophore (× 3)
(*Silene gallica*)
心皮柄（× 3）
（蝇子草）

Bivalved capsule
(*Gentiana* sp.)
两瓣裂的蒴果
（龙胆属某植物）

Other Inflorescence and Flower Features 2
其他花序和花的特点 2

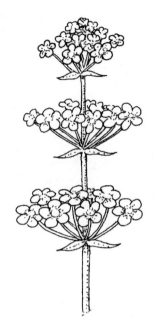

Candelabriform inflorescence
(*Primula japonica*)
烛台状花序（日本报春）

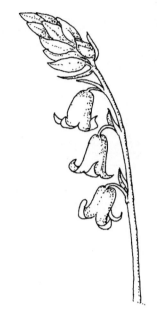

Secund racemose inflorescence
(*Hyacinthoides non-scripta*)
偏向的总状花序（英国蓝铃花）

Diagram of an actinomorphic flower
showing radial symmetry
辐射对称花的示意图，示辐射对称

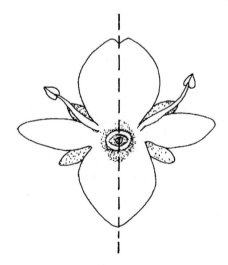

Diagram of a zygomorphic flower
showing bilateral symmetry
两侧对称花的示意图，示两侧对称

Calyces and Epicalyces
花萼和副萼

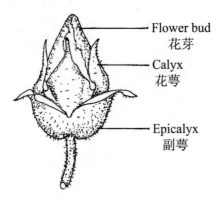

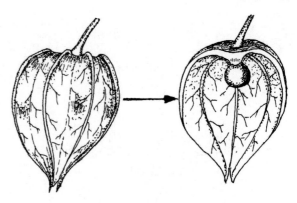

Calyx amd epicalyx of a flower bud
(*Lavatera arborea*)
花芽的花萼和副萼（树状花葵）

Inflated calyx whole and in section
(*Physalis alkekengi*)
膨大花萼的整体和纵切（酸浆）

Flower Forms 1
花的形状 1

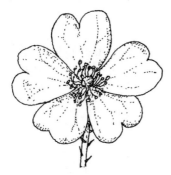

Single flower (*Rosa* sp.)
单花（蔷薇属某植物）

Semi–double flower (*Rosa* hybrid)
半重瓣花（杂种香水月季）

Double flower (*Rosa* hybrid)
重瓣花（杂种香水月季）

Globular flower (*Trollius europaeus*)
球形花（欧洲金莲花）

Flower Forms 2
花的形状 2

Cruciform
十字形的

Papilionaceous
蝶形的

Liliaceous
百合花的

Orchidaceous
兰花的

Campanulate
钟形的

Funnel-shaped
漏斗形的

Rotate
辐射状的

Tubular
筒状的

Urceolate
坛状的

Salver-shaped
高脚碟形的

Personate with spur
具距假面状的

Saccate
囊状的

Labiate
唇形的

Ligulate
舌状的

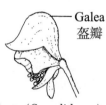
Galea
盔瓣
Galeate (*Consolida* sp.)
盔状的（飞燕草属植物）

Trumpet-shaped
喇叭状的

Gibbous (*Nematanthus gregarius*)
囊状突起的
（袋鼠花）

Dissimilar segments
(*Iris* sp.)
不同裂片的
（鸢尾属植物）

Sepals petaloid
(*Anemone* sp.)
花萼瓣状
（银莲花属植物）

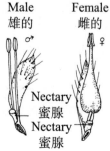
Male
雄的 ♂
Female
雌的 ♀
Nectary
蜜腺
Nectary
蜜腺
Achlamydeous (*Salix* sp.)
无花被的
（柳属植物）

Pollen
花粉

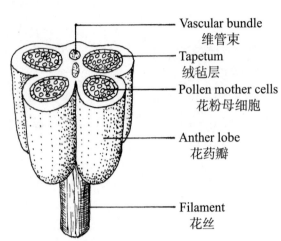

Vascular bundle
维管束

Tapetum
绒毡层

Pollen mother cells
花粉母细胞

Anther lobe
花药瓣

Filament
花丝

Portion of anther before dehiscence
开裂前的花药部分

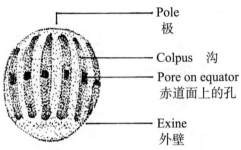

Pole
极

Colpus　沟

Pore on equator
赤道面上的孔

Exine
外壁

Pollen grain of *Polygala chamaebuxus*
荷包山桂花的花粉粒

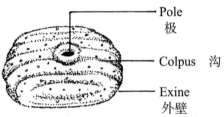

Pole
极

Colpus　沟

Exine
外壁

Pollen grain of *Mimulopsis solmsii*
邵氏拟沟酸浆的花粉粒

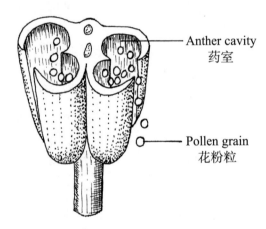

Anther cavity
药室

Pollen grain
花粉粒

Portion of anther after dehiscence
开裂后的花药部分

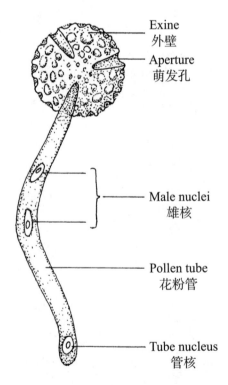

Exine
外壁

Aperture
萌发孔

Male nuclei
雄核

Pollen tube
花粉管

Tube nucleus
管核

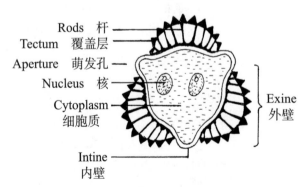

Rods　杆

Tectum　覆盖层

Aperture　萌发孔

Nucleus　核

Cytoplasm
细胞质

Intine
内壁

Exine
外壁

Pollen grain structure
花粉粒的结构

Germination of a pollen grain
花粉粒的萌发

Angiosperm Fertilisation
被子植物的受精

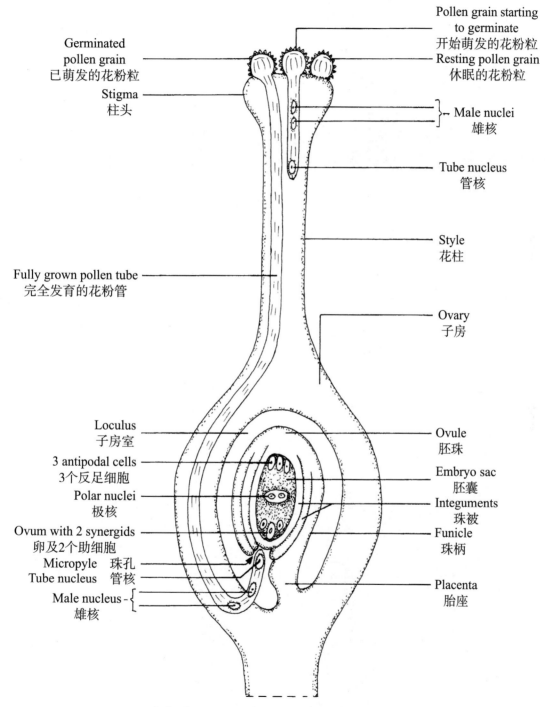

Pollen grain starting
to germinate
开始萌发的花粉粒

Germinated
pollen grain
已萌发的花粉粒

Resting pollen grain
休眠的花粉粒

Stigma
柱头

Male nuclei
雄核

Tube nucleus
管核

Style
花柱

Fully grown pollen tube
完全发育的花粉管

Ovary
子房

Loculus
子房室

Ovule
胚珠

3 antipodal cells
3个反足细胞

Embryo sac
胚囊

Polar nuclei
极核

Integuments
珠被

Ovum with 2 synergids
卵及2个助细胞

Funicle
珠柄

Micropyle　珠孔
Tube nucleus　管核

Male nucleus -
雄核

Placenta
胎座

L.S. of gynoecium showing fertilisation process
雌蕊的纵切，示受精过程

Stamen Arrangement 1
雄蕊的排列　1

Antisepalous
对萼的

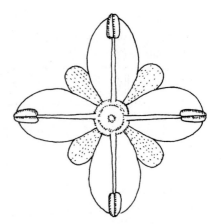

Antipetalous
对瓣的

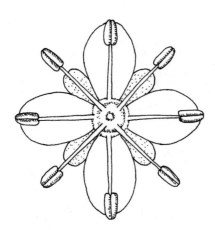

Obdiplostemonous
具外轮对瓣雄蕊的

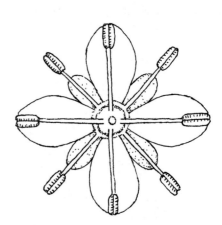

Diplostemonous
具外轮对萼雄蕊的

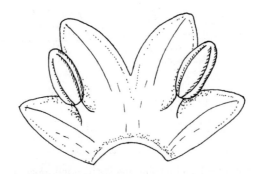

Epipetalous
花冠上着生的

Stamen Arrangement 2
雄蕊的排列 2

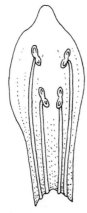

Didynamous (Labiatae or Lamiaceae)
二强雄蕊的（唇形科）

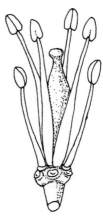

Tetradynamous (Cruciferae or Brassicaceae)
四强雄蕊的（十字花科）

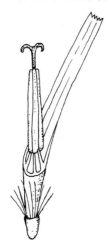

Syngenesious (Compositae or Asteraceae)
聚药雄蕊的（菊科）

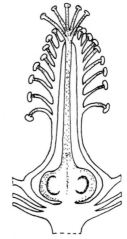

Monadelphous (Malvaceae)
单体雄蕊的（锦葵科）

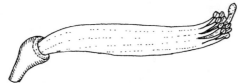

Monadelphous (Leguminosae or Fabaceae)
单体雄蕊的（豆科或蝶形花科）

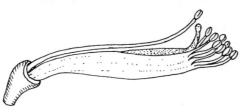

Diadelphous (Leguminosae or Fabaceae)
二体雄蕊的（豆科或蝶形花科）

Stamen Types 1
雄蕊的类型 1

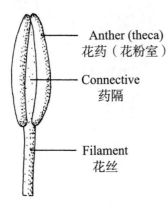

Anther (theca)
花药（花粉室）

Connective
药隔

Filament
花丝

Basifixed (dorsal view)
基着的（背面观）

Basifixed (ventral view)
基着的（腹面观）

Dorsifixed
背着的

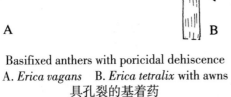

Awn
芒

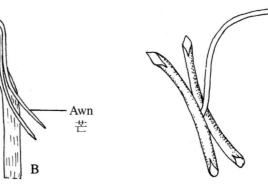

Basifixed anthers with poricidal dehiscence
A. *Erica vagans*   B. *Erica tetralix* with awns
具孔裂的基着药
A. 漂泊欧石楠   B. 具芒的欧石楠

Dorsifixed (versatile)
背着的（丁字着药的）

Anthers of *Berberis darwinii* with valvate dehiscence
A. Before dehiscence   B. After dehiscence
具瓣裂的达尔文小檗的花药
A. 开裂前   B. 开裂后

Petaloid stamen of *Nymphaea*
sp. showing broad connective
睡莲属某植物花瓣状雄
蕊，示宽的药隔

Stamen Types 2
雄蕊的类型 2

Anthers with nectar spurs
(*Viola* sp.)
具蜜距的花药
（堇菜属某植物）

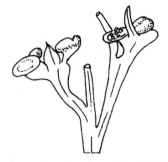

Several anthers on a branched
filament (*Ricinus communis*)
分枝花丝上的几枚花药
（篦麻）

Ventral view   Dorsal view
腹面观   背面观

Anthers with transverse
attachment (*Russelia* sp.)
横向着生的花药
（炮杖属某植物）

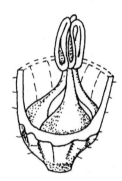

Anthers coherent, arching over
nectary (*Cucurbita* sp.)
蜜腺上方的粘合的弓形花药
（南瓜属某植物）

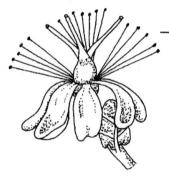 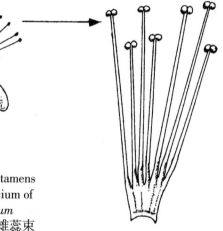

Bundles or fascicles of stamens
surrounding the gynoecium of
Hypericum perforatum
贯叶连翘雌蕊周围的雄蕊束

Fascicle of stamens connate
near the base
近基部合生的雄蕊束

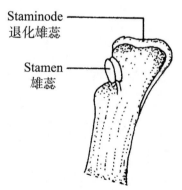

Staminode
退化雄蕊

Stamen
雄蕊

Fertile stamen with single cell attached to a
staminode (*Calathea* sp.)
附着在退化雄蕊上的具有单室的可育雄蕊
（肖竹芋属某植物）

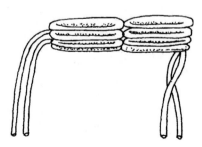

Connate anthers of *Columnea* x *banksii*
飞鱼草（新拟）的合生花药

Dehiscence of Stamen
雄蕊的开裂

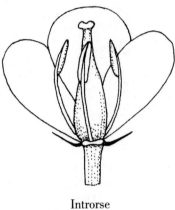

Introrse
内向的

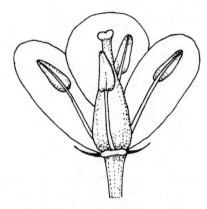

Extrorse
外向的

Ovary Position
子房的位置

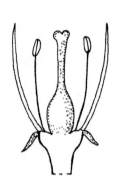

Ovary superior, flower hypogynous
子房上位，花下位

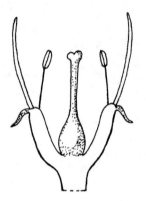

Ovary superior, flower perigynous
子房上位，花周位

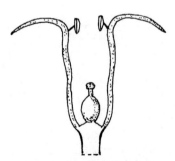

Ovary superior, flower perigynous
子房上位，花周位

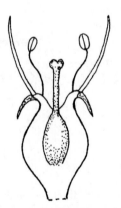

Ovary inferior, flower epigynous
子房下位，花上位

Ovaries and Carpels
子房和心皮

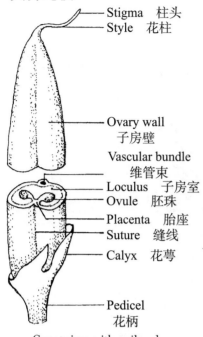

Stigma 柱头
Style 花柱
Ovary wall 子房壁
Vascular bundle 维管束
Loculus 子房室
Ovule 胚珠
Placenta 胎座
Suture 缝线
Calyx 花萼
Pedicel 花柄

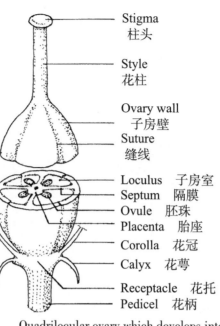

Stigma 柱头
Style 花柱
Ovary wall 子房壁
Suture 缝线
Loculus 子房室
Septum 隔膜
Ovule 胚珠
Placenta 胎座
Corolla 花冠
Calyx 花萼
Receptacle 花托
Pedicel 花柄

Septifragal
室轴开裂的

Loculicidal
室背开裂的

Septicidal
室间开裂的

Gynoecium with unilocular ovary
(legume of *Vicia faba*)
具单室子房的雌蕊（蚕豆的荚果）

Quadrilocular ovary which develops into
a capsule (*Euonymus europaeus*)
发育成蒴果的四室子房（欧洲卫矛）

CARPELS FREE (APOCARPOUS)　心皮离生（离生心皮的）

Several free carpels
(*Aconitum* sp.)
几枚离生心皮
（乌头属某植物）

One free carpels
(*Lathyrus* sp.)
一枚离生心皮
（山黧豆属某植物）

Several free carpels
(*Geum* sp.)
几枚离生心皮
（路边青属某植物）

5 united carpels
surmounted by 5 free
styles (*Cerastium* sp.)
顶端冠以5枚离生
花柱的合生5心皮
（卷耳属某植物）

3-branched style
具3分枝的花柱

3 united carpels
surmounted by a single
style (*Juncus* sp.)
顶端冠以单枚花柱的
合生3心皮
（灯心草属某植物）

5 united carpels surmounted
by an unbranched style
(*Primula* sp.)
顶端冠以一枚不分枝花
柱的合生5心皮
（报春花属某植物）

2 united carpels
surmounted by a very
short styles (*Arabis* sp.)
顶端冠以一枚极短花
柱的合生2心皮
（南芥属某植物）

Styles and Stigmas 1
花柱和柱头 1

Style eccentric
(*Ranunculus* sp.)
花柱离心
（毛茛属某植物）

Style flabellate
(*Viola* sp.)
花柱扇形
（堇菜属某植物）

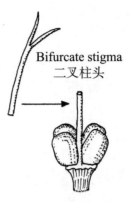

Bifurcate stigma
二叉柱头

Style gynobasic and stigma
bifurcate (*Lamium* sp.)
花柱基着和柱头二叉
（野芝麻属某植物）

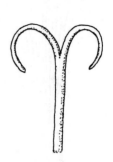

Stigma decurrent
柱头下延

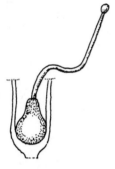

Style geniculate
(*Rhexia* sp.)
花柱膝曲
（鹿草属某植物）

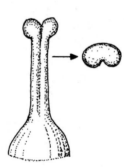

Style conduplicate
花柱对折

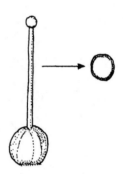

Style terete (*Primula* sp.)
花柱圆柱状
（报春花属某植物）

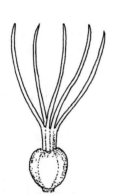

Style with filiform branches
(*Armeria* sp.)
花柱具丝状分枝
（海石竹属某植物）

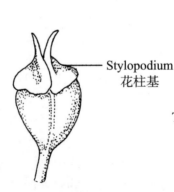

Stylopodium
花柱基

Styles stylopolic
(*Heracleum* sp.)
花柱具花柱基
（独活属某植物）

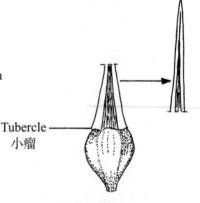

Tubercle
小瘤

Style tuberculate
(*Rhynchospora* sp.)
花柱具瘤
（刺子莞属某植物）

Styles and Stigmas 2
花柱和柱头 2

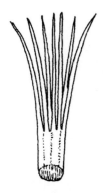

Styles connate
(*Malva* sp.)
花柱合生
（锦葵属某植物）

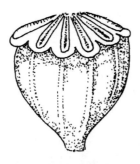

Stigmas sessile and
radiate (*Papaver* sp.)
柱头无柄而呈辐射状
（罂粟属某植物）

Stigma capitate
(*Primula* sp.)
柱头头状
（报春花属某植物）

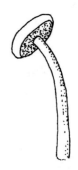

Stigma discoid
(*Hibiscus* sp.)
柱头盘状
（木槿属某植物）

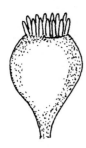

Stigmas forming a crest
柱头形成鸡冠状突起

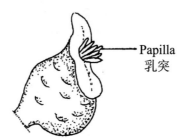

Papilla
乳突

Stigmas sessile with
papillae (*Peperomia* sp.)
柱头无柄，具乳突的
（草胡椒属某植物）

Stigmas fimbriate
柱头流苏状

Stigmas plumose
(Gramineae or Poaceae)
柱头羽毛状（禾本科）

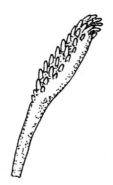

Stigma linear
(*Kitaibelia* sp.)
柱头条形
（凯泰葵属某植物）

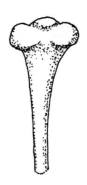

Stigma lobed
(*Lilium* sp.)
柱头浅裂（百合属某植物）

Placentation 1
胎座 1

AXILE
中轴的

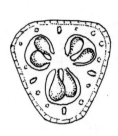

T.S. and L.S. of ovary (*Galanthus nivalis*)
子房横切和纵切（雪花莲）

PARIETAL
侧膜的

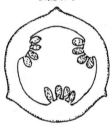

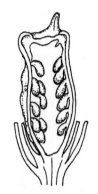

T.S. and L.S. of ovary (*Reseda lutea*)
子房横切和纵切（黄木樨草）

FREE-CENTRAL
特立中央的

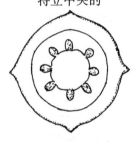
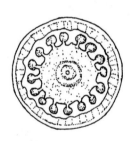
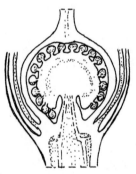

T.S. and L.S. of ovary (*Primula vulgaris*)
子房横切和纵切（欧洲报春）

MARGINAL
边缘的

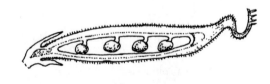

T.S. and L.S. of ovary (*Vicia faba*) showing single row of ovules along a marginal placenta
子房（蚕豆）横切和纵切，示沿边缘胎座的单列胚珠

Placentation 2
胎座 2

MARGINAL
边缘的

  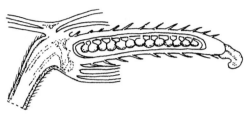

T.S. and L.S. of ovary of *Consolida ajacis* showing numerous ovules on a marginal placenta
飞燕草子房横切和纵切，示边缘胎座上的多数胚珠

APICAL
顶生的

BASAL
基生的

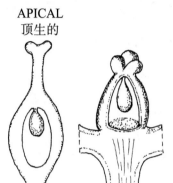 

L.S. of ovary of *Cotinus coggygria*
黄栌子房纵切

L.S. of ovary of *Polygonum persicaria*
春蓼子房纵切

Ovules
胚珠

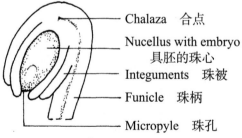

— Chalaza　合点
— Nucellus with embryo
　具胚的珠心
— Integuments　珠被
— Funicle　珠柄
— Micropyle　珠孔

Anatropous
倒生的

Orthotropous
直生的

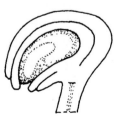

Campylotropous
弯生的

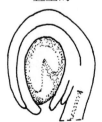

Amphitropous
横生的

10. Flower Structure
10. 花的结构

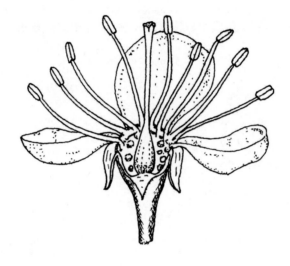

Flower Structure 1 *Aristolochia* and *Sarracenia*
花的结构 1 马兜铃属和瓶子草属

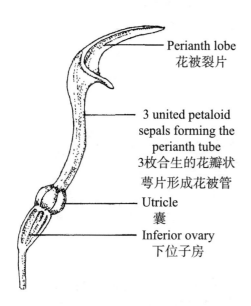

Perianth lobe
花被裂片

3 united petaloid
sepals forming the
perianth tube
3枚合生的花瓣状
萼片形成花被管

Utricle
囊

Inferior ovary
下位子房

Flower of *Aristolochia clematitis*
铁线莲状马兜铃的花

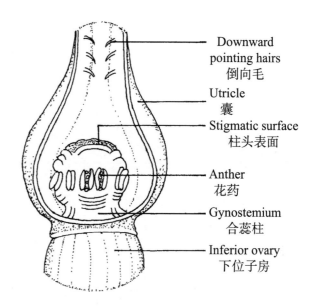

Downward
pointing hairs
倒向毛

Utricle
囊

Stigmatic surface
柱头表面

Anther
花药

Gynostemium
合蕊柱

Inferior ovary
下位子房

L.S. of utricle showing details of the gynostemium
囊的纵切，示合蕊柱的详图

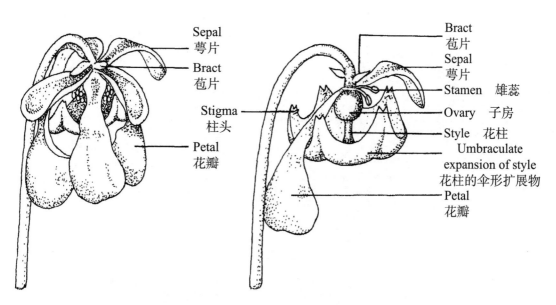

Sepal
萼片

Bract
苞片

Stigma
柱头

Petal
花瓣

Flower of *Sarracenia* sp.
瓶子草属某植物的花

Bract
苞片

Sepal
萼片

Stamen 雄蕊

Ovary 子房

Style 花柱

Umbraculate
expansion of style
花柱的伞形扩展物

Petal
花瓣

Flower of *Sarracenia* sp. with some sepals and
petals removed to reveal the reproductive parts
剥去某些萼片和花瓣露出生殖部分的瓶子草
属某植物的花

Flower Structure 2   *Consolida* and *Salvia*
花的结构 2   飞燕草属和鼠尾草属

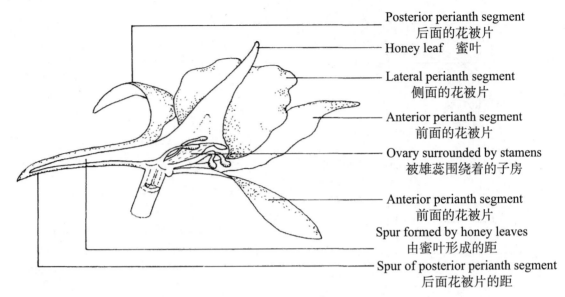

Posterior perianth segment
后面的花被片
Honey leaf　蜜叶
Lateral perianth segment
侧面的花被片
Anterior perianth segment
前面的花被片
Ovary surrounded by stamens
被雄蕊围绕着的子房
Anterior perianth segment
前面的花被片
Spur formed by honey leaves
由蜜叶形成的距
Spur of posterior perianth segment
后面花被片的距

L.S. of flower of *Consolida ajacis*, a zygomorphic flower with separate perianth segments
飞燕草花的纵切，具离生花被片的两侧对称花

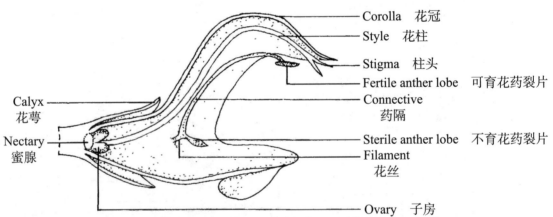

Corolla　花冠
Style　花柱
Stigma　柱头
Fertile anther lobe　可育花药裂片
Connective
药隔
Calyx
花萼
Sterile anther lobe　不育花药裂片
Filament
花丝
Nectary
蜜腺
Ovary　子房

L.S. of flower of *Salvia pratensis*, a zygomorphic flower with united perianth segments
草原鼠尾草花的纵切，具合生花被片的两侧对称花

Fertile anther lobe　可育花药裂片
Connectives of the paired stamens
成对雄蕊的药隔
The 2 joined anther lobes　2枚连接的花药裂片
Filament attached to corolla tube　花丝着生到花冠筒上

Detail of the paired stamens of *Salvia pratensis*
草原鼠尾草成对雄蕊的详图

Flower Structure 3   *Silene dioica*
花的结构 3   异株蝇子草

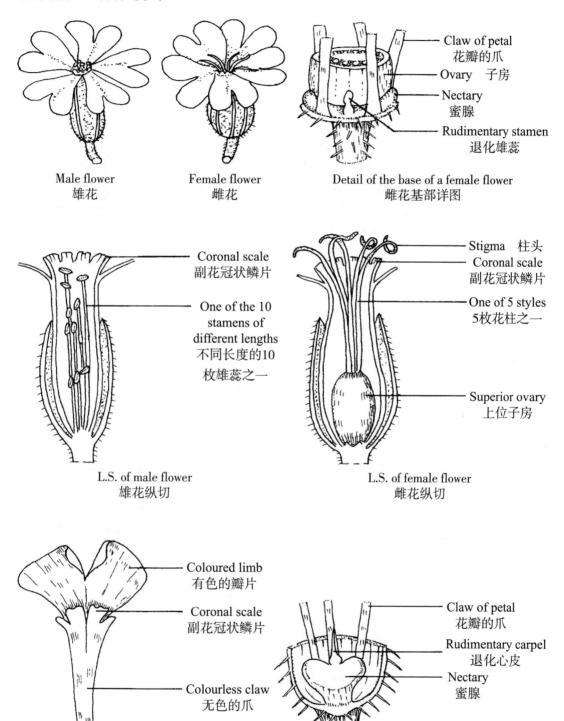

Male flower
雄花

Female flower
雌花

Detail of the base of a female flower
雌花基部详图

Claw of petal
花瓣的爪

Ovary   子房

Nectary
蜜腺

Rudimentary stamen
退化雄蕊

Coronal scale
副花冠状鳞片

One of the 10
stamens of
different lengths
不同长度的10
枚雄蕊之一

Stigma   柱头

Coronal scale
副花冠状鳞片

One of 5 styles
5枚花柱之一

Superior ovary
上位子房

L.S. of male flower
雄花纵切

L.S. of female flower
雌花纵切

Coloured limb
有色的瓣片

Coronal scale
副花冠状鳞片

Colourless claw
无色的爪

Claw of petal
花瓣的爪

Rudimentary carpel
退化心皮

Nectary
蜜腺

Detail of unguiculate petal
具爪花瓣详图

Detail of the base of a male flower
雄花基部详图

Flower Structure 4 *Passiflora*
花的结构 4 西番莲属

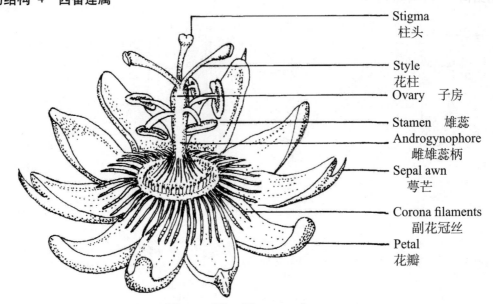

Stigma
柱头

Style
花柱
Ovary 子房

Stamen 雄蕊
Androgynophore
雌雄蕊柄
Sepal awn
萼芒

Corona filaments
副花冠丝
Petal
花瓣

Flower of *Passiflora caerulea*
蓝花西番莲的花

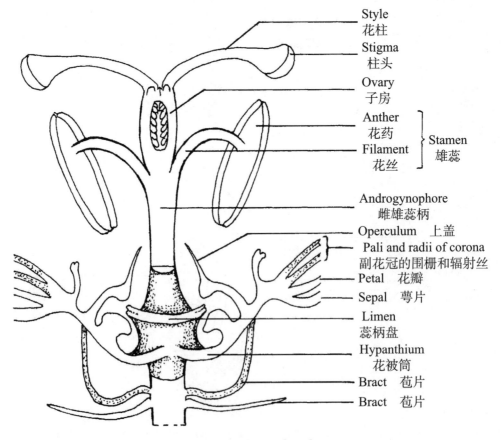

Style
花柱
Stigma
柱头

Ovary
子房

Anther
花药
\} Stamen
雄蕊
Filament
花丝

Androgynophore
雌雄蕊柄
Operculum 上盖
Pali and radii of corona
副花冠的围栅和辐射丝
Petal 花瓣
Sepal 萼片
Limen
蕊柄盘
Hypanthium
花被筒
Bract 苞片
Bract 苞片

L.S. of portion of *Passiflora* flower
西番莲花部纵切

Flower Structure 5　*Primula, Cyclamen* and *Viola*
花的结构 5　报春花属，仙客来属和堇菜属

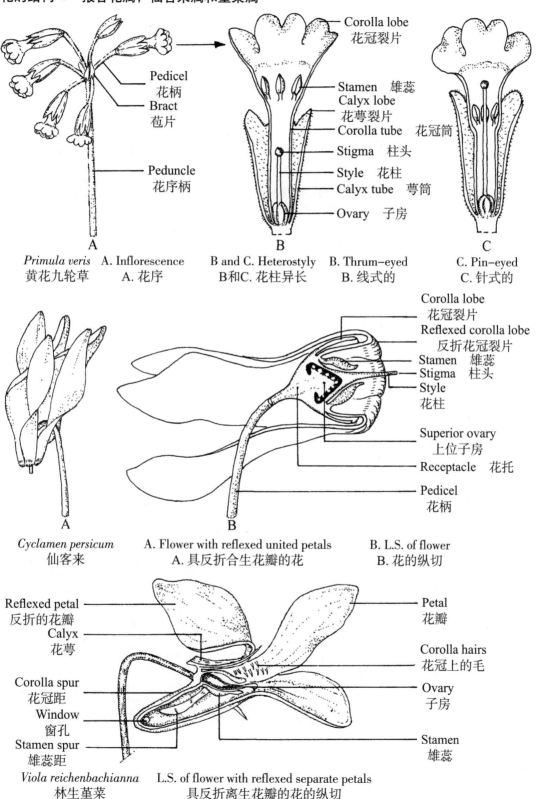

Corolla lobe
花冠裂片

Pedicel
花柄
Bract
苞片

Stamen　雄蕊
Calyx lobe
花萼裂片
Corolla tube　花冠筒
Stigma　柱头
Style　花柱
Calyx tube　萼筒
Ovary　子房

Peduncle
花序柄

A

B

C

Primula veris　A. Inflorescence
黄花九轮草　　A. 花序

B and C. Heterostyly　B. Thrum-eyed
B和C. 花柱异长　　B. 线式的

C. Pin-eyed
C. 针式的

Corolla lobe
花冠裂片
Reflexed corolla lobe
反折花冠裂片
Stamen　雄蕊
Stigma　柱头
Style
花柱

Superior ovary
上位子房

Receptacle　花托

Pedicel
花柄

A

B

Cyclamen persicum　A. Flower with reflexed united petals
仙客来　　　　　A. 具反折合生花瓣的花

B. L.S. of flower
B. 花的纵切

Reflexed petal
反折的花瓣
Calyx
花萼

Corolla spur
花冠距
Window
窗孔
Stamen spur
雄蕊距

Petal
花瓣

Corolla hairs
花冠上的毛
Ovary
子房

Stamen
雄蕊

Viola reichenbachianna　L.S. of flower with reflexed separate petals
林生堇菜　　　　　具反折离生花瓣的花的纵切

Flower Structure 6 *Lathyrus*

花的结构 6 山黧豆属

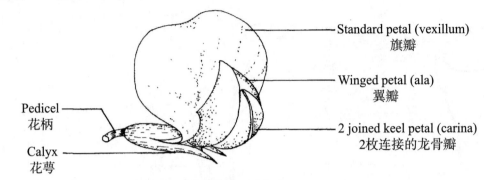

Standard petal (vexillum)
旗瓣

Winged petal (ala)
翼瓣

2 joined keel petal (carina)
2枚连接的龙骨瓣

Pedicel
花柄

Calyx
花萼

Zygomorphic flower of *Lathyrus* sp.
山黧豆属某植物的两侧对称花

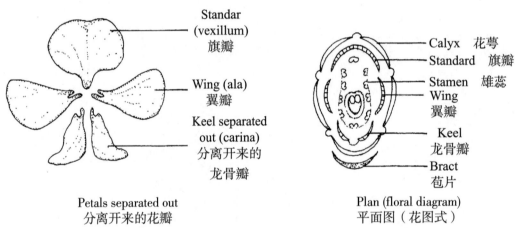

Standar
(vexillum)
旗瓣

Wing (ala)
翼瓣

Keel separated
out (carina)
分离开来的
龙骨瓣

Petals separated out
分离开来的花瓣

Calyx 花萼
Standard 旗瓣
Stamen 雄蕊
Wing
翼瓣
Keel
龙骨瓣
Bract
苞片

Plan (floral diagram)
平面图（花图式）

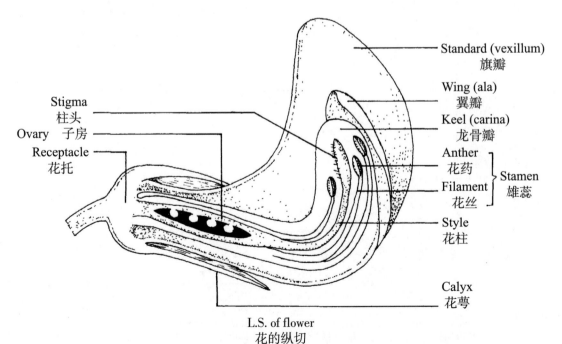

Standard (vexillum)
旗瓣

Wing (ala)
翼瓣

Keel (carina)
龙骨瓣

Anther
花药
Filament
花丝

Stamen
雄蕊

Style
花柱

Calyx
花萼

Stigma
柱头
Ovary 子房
Receptacle
花托

L.S. of flower
花的纵切

Flower Structure 7 *Eucalyptus,* **Nectar Guides and Nectaries**
花的结构 7 桉属，花蜜指示标和蜜腺

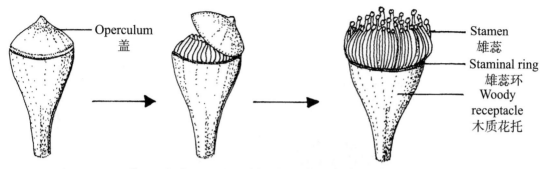

Operculum
盖

Stamen
雄蕊

Staminal ring
雄蕊环

Woody
receptacle
木质花托

Stages in the opening of the flower bud in *Eucalyptus* sp.
桉属某植物的花芽开放期

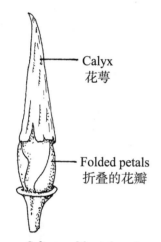

Calyx
花萼

Folded petals
折叠的花瓣

Calyptra–like calyx of
Eschscholzia californica
花菱草的帽状花萼

Nectar guides on the
lip of *Ajuga reptans*
匍匐筋骨草唇瓣
上的花蜜指示标

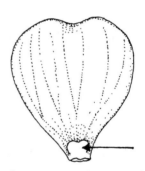

Pocket nectary at the petal base
of *Ranunculus repens*
匍枝毛茛花瓣基部的囊状蜜腺

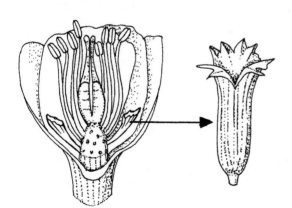

Petal nectary *in situ* and separated out from the
flower of *Helleborus foetidus*
臭圣诞玫瑰原始位置的和分离出来的花瓣蜜腺

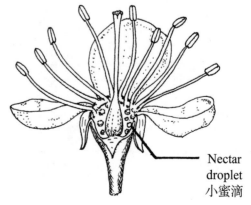

Nectar
droplet
小蜜滴

Nectar–secreting hypanthium of the flower of
Prunus spinosa
黑刺李花的分泌蜜汁的花筒

Flower Structure 8 *Geranium*
花的结构 8 老鹳草属

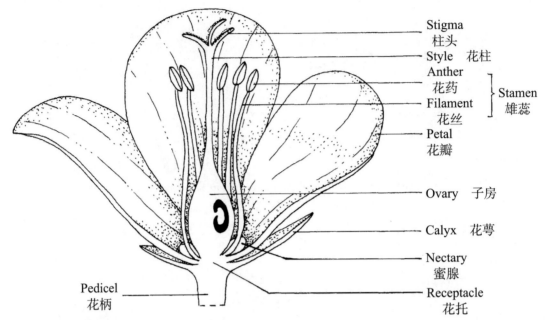

Stigma
柱头

Style 花柱

Anther
花药

Filament
花丝
} Stamen
雄蕊

Petal
花瓣

Ovary 子房

Calyx 花萼

Nectary
蜜腺

Pedicel
花柄

Receptacle
花托

L.S. of a *Geranium*, a flower with 5 free sepals, 5 free petals, and a superior ovary
老鹳草属植物一朵具5枚离生萼片、5枚离生花瓣和1枚上位子房的花的纵切

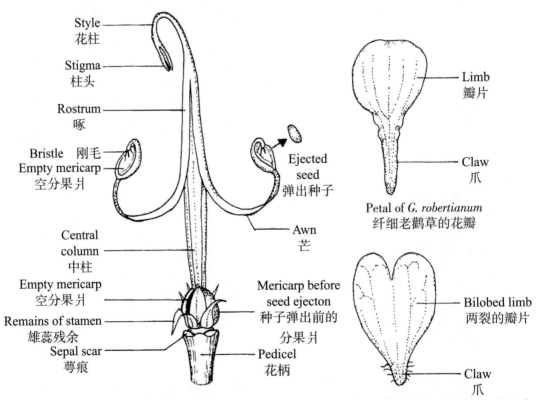

Style
花柱

Stigma
柱头

Rostrum
啄

Bristle 刚毛
Empty mericarp
空分果爿

Central
column
中柱

Empty mericarp
空分果爿

Remains of stamen
雄蕊残余

Sepal scar
萼痕

Ejected
seed
弹出种子

Awn
芒

Mericarp before
seed ejecton
种子弹出前的
分果爿

Pedicel
花柄

Seed dispersal in a *Geranium* sp.
老鹳草属某植物的种子散布

Limb
瓣片

Claw
爪

Petal of *G. robertianum*
纤细老鹳草的花瓣

Bilobed limb
两裂的瓣片

Claw
爪

Petal of *G. molle* showing structure of petals
柔老鹳草的花瓣，示花瓣的结构

Flower Structure 9 *Asclepias* and *Polygala*
花的结构 9 马利筋属和远志属

ASCLEPIAS 马利筋属

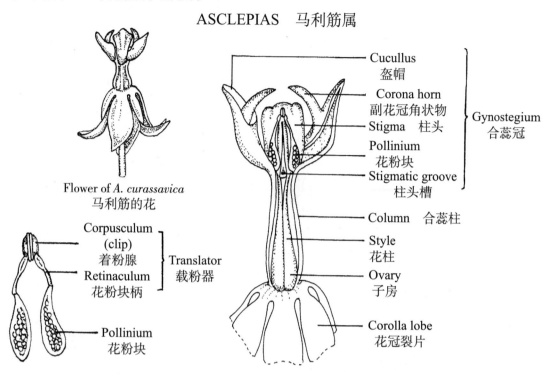

Cucullus
盔帽

Corona horn
副花冠角状物

Stigma 柱头

Pollinium
花粉块

Stigmatic groove
柱头槽

Gynostegium
合蕊冠

Column 合蕊柱

Style
花柱

Ovary
子房

Corolla lobe
花冠裂片

Flower of *A. curassavica*
马利筋的花

Corpusculum
(clip)
着粉腺

Retinaculum
花粉块柄

Translator
载粉器

Pollinium
花粉块

Translator and pair of pollinia
载粉器和成对的花粉块

Corolla, column and corona of *A. curassavica*
马利筋花冠、合蕊柱和副花冠

POLYGALA 远志属

Fimbriated
crest
流苏状的
鸡冠突起

Hook-like stigma
钩状柱头

Spathulate end of style
花柱匙形端

Style
花柱

Flower of *P. vulgaris* with 2 of the petaloid sepals
removed to reveal the anterior petal with its fimbriated
crest and the 2 lateral (upper) petals
去掉2枚花瓣状萼片，露出具流苏状鸡冠状突起的
前花瓣和两枚侧生（上面）的花瓣的普通远志的花

Upper portion of spathulate style and stigma
of *P. vulgaris*
普通远志的匙形花柱上部和柱头

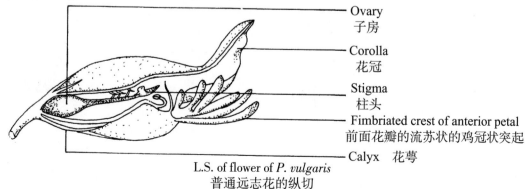

Ovary
子房

Corolla
花冠

Stigma
柱头

Fimbriated crest of anterior petal
前面花瓣的流苏状的鸡冠状突起

Calyx 花萼

L.S. of flower of *P. vulgaris*
普通远志花的纵切

Flower Structure 10   Compositae (Asteraceae) 1
花的结构 10   菊科 1

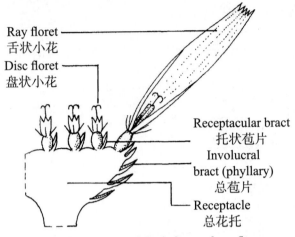

Ray floret
舌状小花

Disc floret
盘状小花

Receptacular bract
托状苞片

Involucral
bract (phyllary)
总苞片

Receptacle
总花托

Portion of capitulum with both disc and ray florets
具盘状小花和舌状小花的头状花序部分

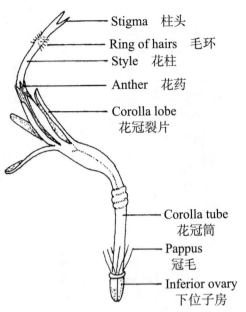

Stigma   柱头

Ring of hairs   毛环

Style   花柱

Anther   花药

Corolla lobe
花冠裂片

Corolla tube
花冠筒

Pappus
冠毛

Inferior ovary
下位子房

Tubular floret of *Centaurea* sp.
矢车菊属某植物的筒状小花

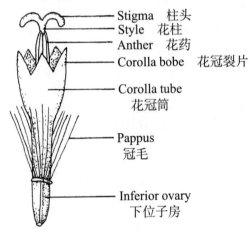

Stigma   柱头
Style   花柱
Anther   花药
Corolla bobe   花冠裂片

Corolla tube
花冠筒

Pappus
冠毛

Inferior ovary
下位子房

Disc floret of *Doronicum* sp.
多榔菊属某植物的盘状小花

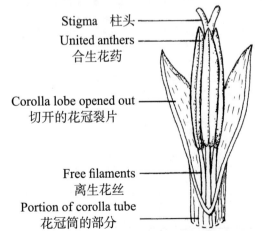

Stigma   柱头
United anthers
合生花药

Corolla lobe opened out
切开的花冠裂片

Free filaments
离生花丝

Portion of corolla tube
花冠筒的部分

Disc floret of *Doronicum* sp. showing detail of stamens
多榔菊属某植物的盘状小花，示雄蕊的详图

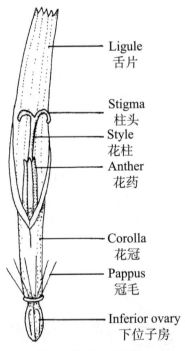

Ligule
舌片

Stigma
柱头
Style
花柱
Anther
花药

Corolla
花冠

Pappus
冠毛

Inferior ovary
下位子房

Ray or ligulate floret
舌状小花

Flower Structure 11　Compositae (Asteraceae) 2
花的结构 11　菊科 2

CAPITULA
头状花序

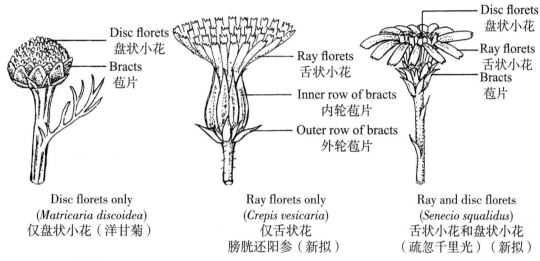

Disc florets
盘状小花

Bracts
苞片

Ray florets
舌状小花

Inner row of bracts
内轮苞片

Outer row of bracts
外轮苞片

Disc florets
盘状小花

Ray florets
舌状小花

Bracts
苞片

Disc florets only
(*Matricaria discoidea*)
仅盘状小花（洋甘菊）

Ray florets only
(*Crepis vesicaria*)
仅舌状花
膀胱还阳参（新拟）

Ray and disc florets
(*Senecio squalidus*)
舌状小花和盘状小花
（疏忽千里光）（新拟）

FRUITS　果实

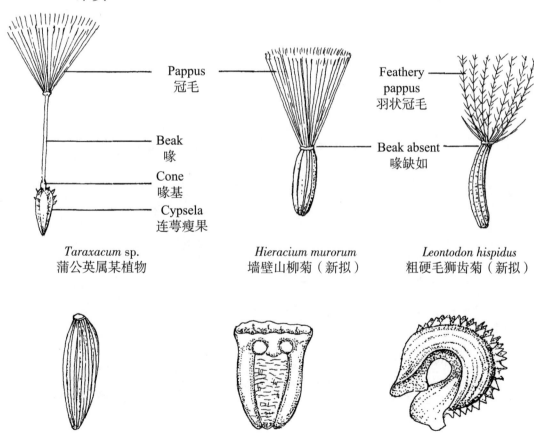

Pappus
冠毛

Beak
喙

Cone
喙基

Cypsela
连萼瘦果

Feathery
pappus
羽状冠毛

Beak absent
喙缺如

Taraxacum sp.
蒲公英属某植物

Hieracium murorum
墙壁山柳菊（新拟）

Leontodon hispidus
粗硬毛狮齿菊（新拟）

Lapsana communis
欧洲倒槎菜

Matricaria recutita
母菊

Calendula arvensis
金盏花

Flower Structure 12　*Strelitzia* and *Canna*
花的结构 12　鹤望兰属和美人蕉属

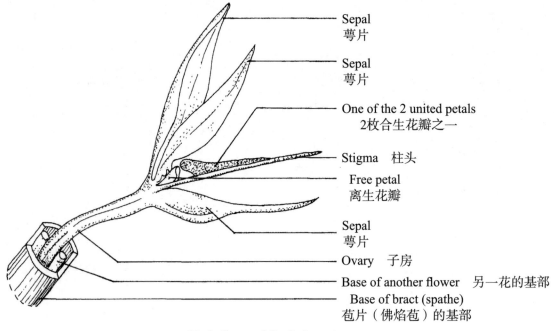

Sepal 萼片

Sepal 萼片

One of the 2 united petals 2枚合生花瓣之一

Stigma　柱头

Free petal 离生花瓣

Sepal 萼片

Ovary　子房

Base of another flower　另一花的基部

Base of bract (spathe) 苞片（佛焰苞）的基部

Single flower of *Strelitzia reginae*
鹤望兰的单花

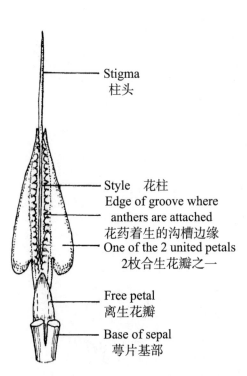

Stigma 柱头

Style　花柱
Edge of groove where anthers are attached
花药着生的沟槽边缘
One of the 2 united petals
2枚合生花瓣之一

Free petal 离生花瓣

Base of sepal 萼片基部

Flower of *S. reginae* with sepals removed
剥去萼片的鹤望兰的花

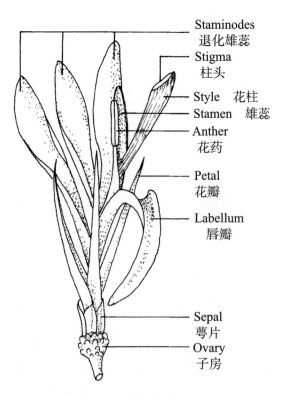

Staminodes 退化雄蕊
Stigma 柱头
Style　花柱
Stamen　雄蕊
Anther 花药
Petal 花瓣
Labellum 唇瓣
Sepal 萼片
Ovary 子房

Flower of *Canna indica*
美人蕉的花

Flower Structure 13 Tulipa and Narcissus
花的结构 13 郁金香属和水仙属

OVARY SUPERIOR 子房上位

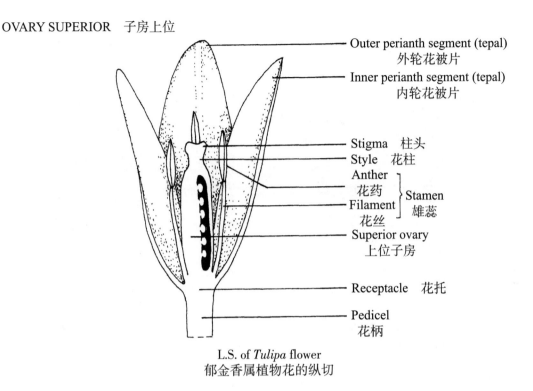

Outer perianth segment (tepal)
外轮花被片
Inner perianth segment (tepal)
内轮花被片

Stigma 柱头
Style 花柱
Anther
花药
Filament } Stamen
花丝 } 雄蕊
Superior ovary
上位子房

Receptacle 花托

Pedicel
花柄

L.S. of *Tulipa* flower
郁金香属植物花的纵切

OVARY INFERIOR 子房下位

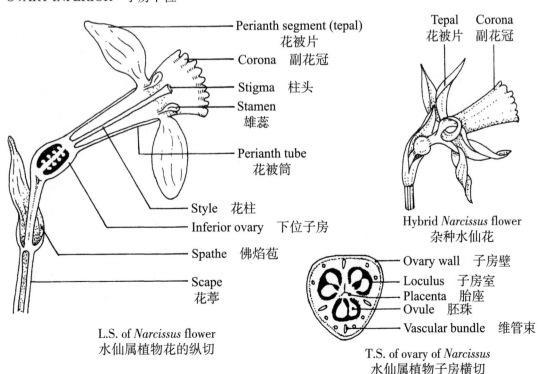

Perianth segment (tepal)
花被片
Corona 副花冠
Stigma 柱头
Stamen
雄蕊

Perianth tube
花被筒

Style 花柱
Inferior ovary 下位子房
Spathe 佛焰苞
Scape
花葶

L.S. of *Narcissus* flower
水仙属植物花的纵切

Tepal   Corona
花被片   副花冠

Hybrid *Narcissus* flower
杂种水仙花

Ovary wall 子房壁
Loculus 子房室
Placenta 胎座
Ovule 胚珠
Vascular bundle 维管束

T.S. of ovary of *Narcissus*
水仙属植物子房横切

Flower Structure 14   *Iris*
花的结构 14 鸢尾属

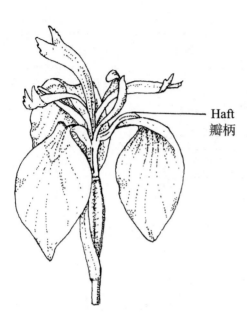

Haft
瓣柄

Flower of *Iris pseudacorus*
黄花鸢尾的花

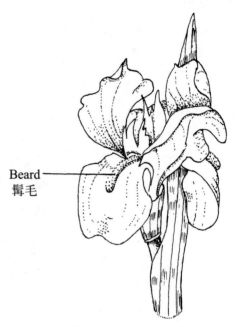

Beard
髯毛

Flower of a bearded *Iris*
具髯毛的鸢尾属植物的花

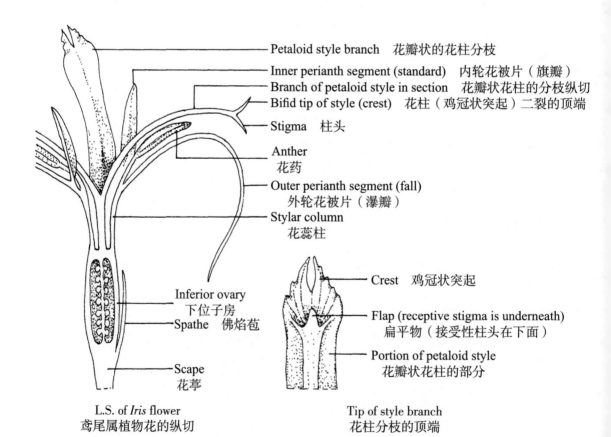

Petaloid style branch   花瓣状的花柱分枝
Inner perianth segment (standard)   内轮花被片（旗瓣）
Branch of petaloid style in section   花瓣状花柱的分枝纵切
Bifid tip of style (crest)   花柱（鸡冠状突起）二裂的顶端
Stigma   柱头
Anther
花药
Outer perianth segment (fall)
外轮花被片（瀑瓣）
Stylar column
花蕊柱

Inferior ovary
下位子房
Spathe   佛焰苞

Scape
花葶

Crest   鸡冠状突起

Flap (receptive stigma is underneath)
扁平物（接受性柱头在下面）

Portion of petaloid style
花瓣状花柱的部分

L.S. of *Iris* flower
鸢尾属植物花的纵切

Tip of style branch
花柱分枝的顶端

11. Features of Certain Plant Families
11. 某些植物科的特征

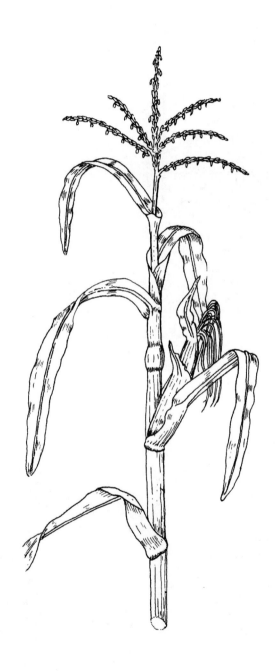

Structure of Cacti
仙人掌科的结构

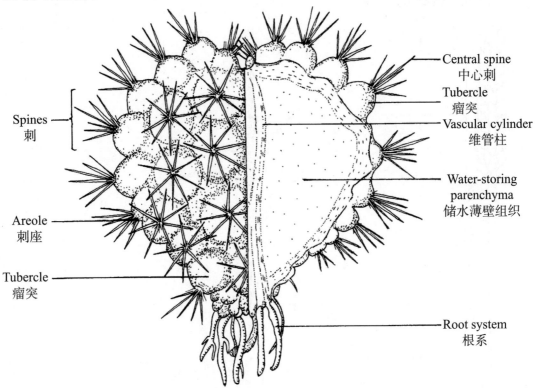

Spines
刺

Areole
刺座

Tubercle
瘤突

Central spine
中心刺

Tubercle
瘤突

Vascular cylinder
维管柱

Water-storing
parenchyma
储水薄壁组织

Root system
根系

Structure of *Echinocactus* sp., whole and in section
仙人球属某植物的结构，整体和纵切

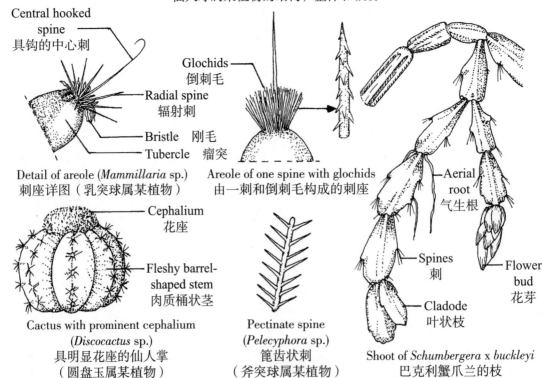

Central hooked
spine
具钩的中心刺

Radial spine
辐射刺

Bristle 刚毛

Tubercle 瘤突

Detail of areole (*Mammillaria* sp.)
刺座详图（乳突球属某植物）

Glochids
倒刺毛

Areole of one spine with glochids
由一刺和倒刺毛构成的刺座

Cephalium
花座

Fleshy barrel-
shaped stem
肉质桶状茎

Cactus with prominent cephalium
(*Discocactus* sp.)
具明显花座的仙人掌
（圆盘玉属某植物）

Pectinate spine
(*Pelecyphora* sp.)
篦齿状刺
（斧突球属某植物）

Aerial
root
气生根

Spines
刺

Cladode
叶状枝

Flower
bud
花芽

Shoot of *Schumbergera* x *buckleyi*
巴克利蟹爪兰的枝

Structure of Grasses 1　　Main Features
禾本科的结构 1　主要特征

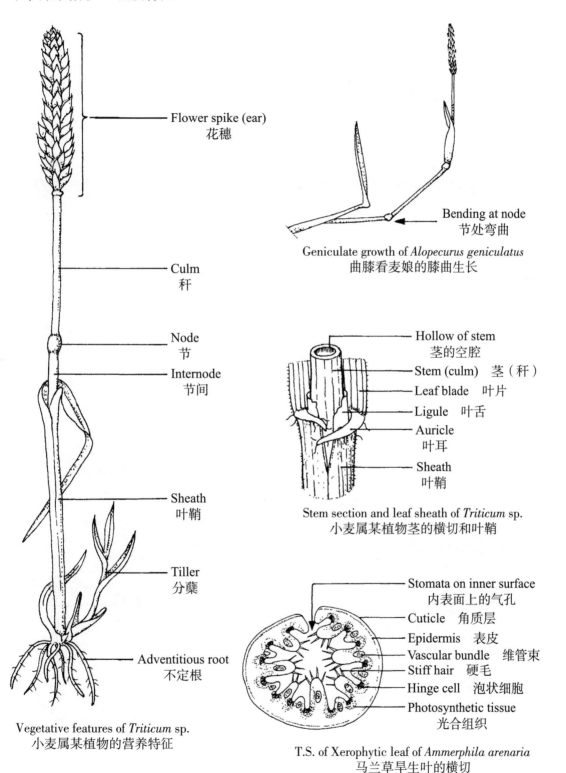

Flower spike (ear)
花穗

Culm
秆

Node
节

Internode
节间

Sheath
叶鞘

Tiller
分蘖

Adventitious root
不定根

Vegetative features of *Triticum* sp.
小麦属某植物的营养特征

Bending at node
节处弯曲

Geniculate growth of *Alopecurus geniculatus*
曲膝看麦娘的膝曲生长

Hollow of stem
茎的空腔
Stem (culm)　茎（秆）
Leaf blade　叶片
Ligule　叶舌
Auricle
叶耳
Sheath
叶鞘

Stem section and leaf sheath of *Triticum* sp.
小麦属某植物茎的横切和叶鞘

Stomata on inner surface
内表面上的气孔
Cuticle　角质层
Epidermis　表皮
Vascular bundle　维管束
Stiff hair　硬毛
Hinge cell　泡状细胞
Photosynthetic tissue
光合组织

T.S. of Xerophytic leaf of *Ammerphila arenaria*
马兰草旱生叶的横切

Structure of Grasses 2   Inflorescence Forms
禾本科的结构 2　花序的形状

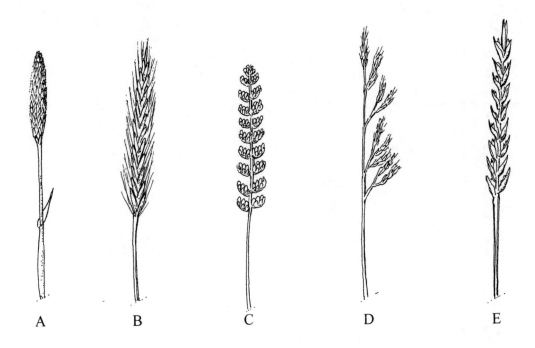

A　　B　　C　　D　　E

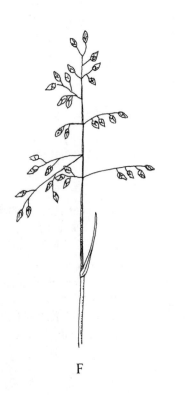

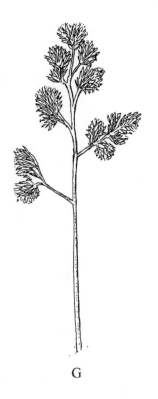

A. *Phleum arenarium*
A. 高山梯牧草
B. *Hordeum murinum*
B. 野大麦
C. *Cynosurus cristatus*
C. 洋狗尾草
D. *Vulpia bromoides*
D. 欧洲福克斯泰尔羊茅
E. *Agropyron pungens*
E. 锐尖冰草
F. *Poa annua*
F. 早熟禾
G. *Dactylis glomerata*
G. 鸭茅

F　　　　　　　G

Structure of Grasses 3
禾本科的结构 3

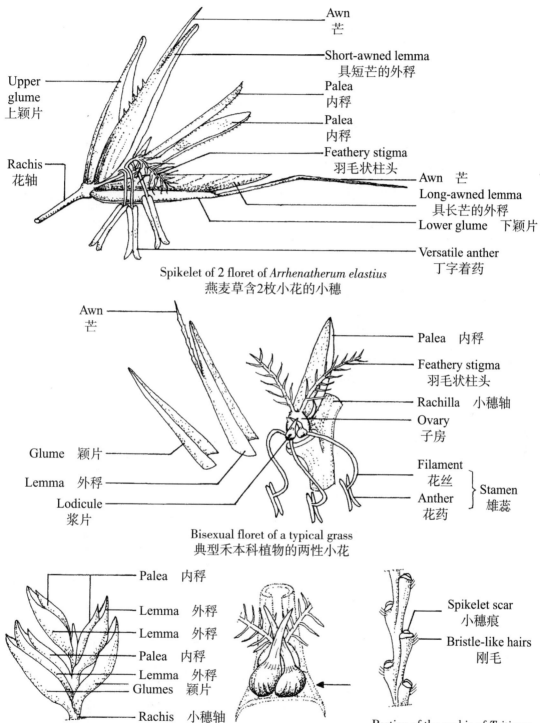

Awn
芒

Short-awned lemma
具短芒的外稃

Palea
内稃

Palea
内稃

Feathery stigma
羽毛状柱头

Awn 芒

Long-awned lemma
具长芒的外稃

Lower glume 下颖片

Versatile anther
丁字着药

Upper
glume
上颖片

Rachis
花轴

Spikelet of 2 floret of *Arrhenatherum elastius*
燕麦草含2枚小花的小穗

Awn
芒

Palea 内稃

Feathery stigma
羽毛状柱头

Rachilla 小穗轴

Ovary
子房

Filament
花丝

Anther
花药

Stamen
雄蕊

Glume 颖片

Lemma 外稃

Lodicule
浆片

Bisexual floret of a typical grass
典型禾本科植物的两性小花

Palea 内稃

Lemma 外稃

Lemma 外稃

Palea 内稃

Lemma 外稃

Glumes 颖片

Rachis 小穗轴

Spikelet containing 3 florets
with 2 glumes at the base
基部具2枚颖片、含3枚
小花的小穗

Two lodicules at the base of the
ovary (*Arrhenatherum elatius*)
子房基部的两个浆片
（燕麦草）

Spikelet scar
小穗痕

Bristle-like hairs
刚毛

Portion of the rachis of *Triticum*
sp. with spikelets removed to
show zig-zag arrangement
除去小穗的小麦属某植物的穗
轴部分，示"之"字形排列

Structure of Grasses 4 *Zea mays*
禾本科的结构 4 玉米

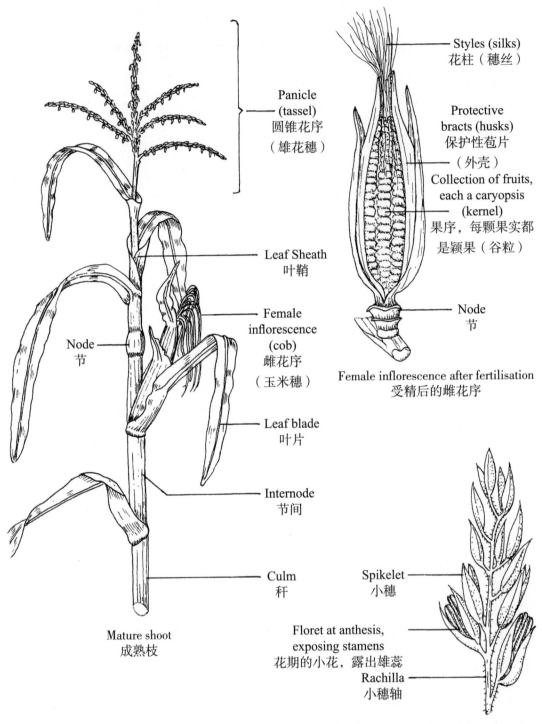

Panicle
(tassel)
圆锥花序
（雄花穗）

Styles (silks)
花柱（穗丝）

Protective
bracts (husks)
保护性苞片
（外壳）

Collection of fruits,
each a caryopsis
(kernel)
果序，每颗果实都
是颖果（谷粒）

Leaf Sheath
叶鞘

Female
inflorescence
(cob)
雌花序
（玉米穗）

Node
节

Node
节

Female inflorescence after fertilisation
受精后的雌花序

Leaf blade
叶片

Internode
节间

Culm
秆

Mature shoot
成熟枝

Spikelet
小穗

Floret at anthesis,
exposing stamens
花期的小花，露出雄蕊
Rachilla
小穗轴

Branch of male inflorescence removed from panicle
从圆锥花序上分出的雄花序的分枝

Structure of Rushes *Juncus* and *Luzula*
灯心草科的结构，灯心草属和地杨梅属

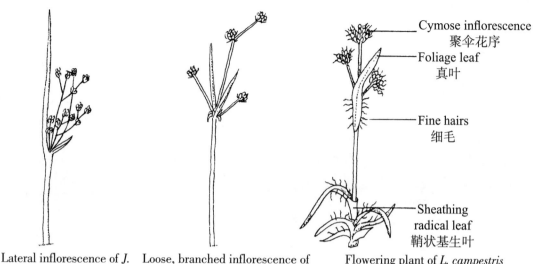

Lateral inflorescence of *J. inflexus*
片髓灯心草的侧生花序

Loose, branched inflorescence of *J. bulbosus*
鳞茎灯心草的松散分枝花序

Cymose inflorescence
聚伞花序

Foliage leaf
真叶

Fine hairs
细毛

Sheathing radical leaf
鞘状基生叶

Flowering plant of *L. campestris*
地杨梅的花期植株

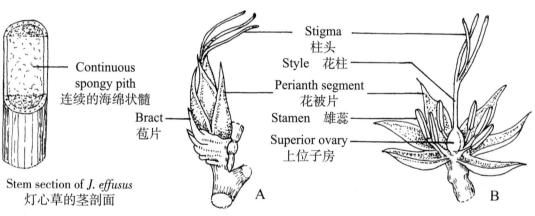

Continuous spongy pith
连续的海绵状髓

Stem section of *J. effusus*
灯心草的茎剖面

Bract
苞片

Flower of *L. campestris*
地杨梅的花

Stigma 柱头
Style 花柱
Perianth segment 花被片
Stamen 雄蕊
Superior ovary 上位子房

A. Bud stage   B. At anthesis
A. 芽期   B. 花期

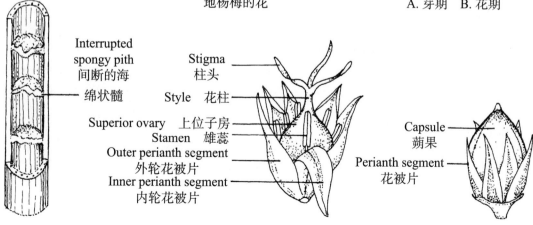

Interrupted spongy pith
间断的海绵状髓

Stem section of *J. inflexus*
片髓灯心草的茎剖面

Stigma 柱头
Style 花柱
Superior ovary 上位子房
Stamen 雄蕊
Outer perianth segment 外轮花被片
Inner perianth segment 内轮花被片

Flower of *J. inflexus*
片髓灯心草的花

Capsule
蒴果

Perianth segment
花被片

Fruit (capsule) of *J. inflexus*
片髓灯心草的果实（蒴果）

Structure of Sedges *Carex*
莎草科的结构，苔草属

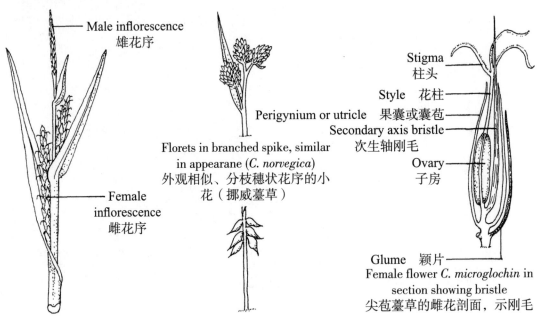

Male inflorescence
雄花序

Female inflorescence
雌花序

Flowering shoot (*C. acutiformis*)
花枝（皱果薹草）

Florets in branched spike, similar in appearane (*C. norvegica*)
外观相似、分枝穗状花序的小花（挪威薹草）

Florets in unbranched spike (*C. microglochin*)
不分枝穗状花序的小花（尖苞薹草）

Stigma
柱头

Style　花柱

Perigynium or utricle　果囊或囊苞

Secondary axis bristle
次生轴刚毛

Ovary
子房

Glume　颖片

Female flower *C. microglochin* in section showing bristle
尖苞薹草的雌花剖面，示刚毛

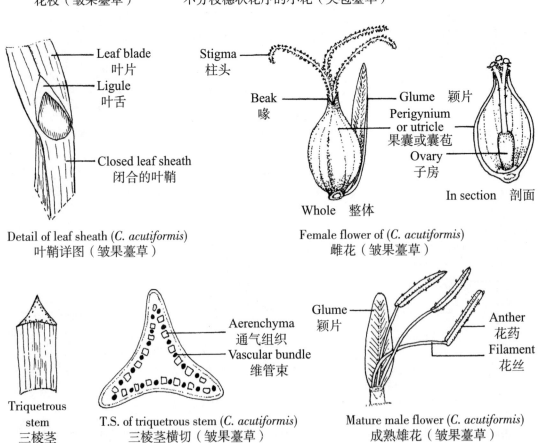

Leaf blade
叶片

Ligule
叶舌

Closed leaf sheath
闭合的叶鞘

Detail of leaf sheath (*C. acutiformis*)
叶鞘详图（皱果薹草）

Stigma
柱头

Beak
喙

Glume　颖片

Perigynium or utricle
果囊或囊苞

Ovary
子房

Whole　整体

In section　剖面

Female flower of (*C. acutiformis*)
雌花（皱果薹草）

Triquetrous stem
三棱茎

Aerenchyma
通气组织

Vascular bundle
维管束

T.S. of triquetrous stem (*C. acutiformis*)
三棱茎横切（皱果薹草）

Glume
颖片

Anther
花药

Filament
花丝

Mature male flower (*C. acutiformis*)
成熟雄花（皱果薹草）

Structure of Palms 1
棕榈科的结构 1

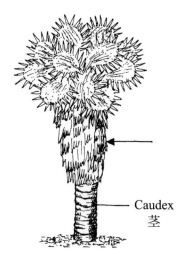

Marcescent leaves (*Washingtonia* sp.)
凋萎的叶（丝葵属某植物）

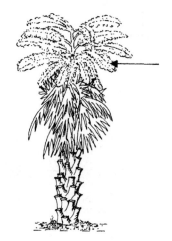

Suprafoliar inflorescence (*Corypha* sp.)
叶上方的花序（贝叶棕属某植物）

Dichotomously branched (*Hyphaene* sp.)
二叉分枝的（叉干棕属某植物）

Acaulescent (*Allagoptera* sp.)
无茎的（刺鱼尾椰属某植物）

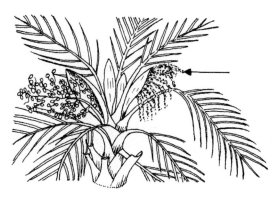

Interfoliar inflorescence (*Butia* sp.)
叶间的花序（椰属某植物）

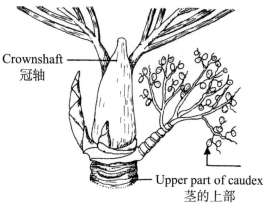

Infrafoliar inflorescence (*Hyophorbe* sp.)
叶下的花序（洒瓶椰子属某植物）

Structure of Palms 2
棕榈科的结构 2

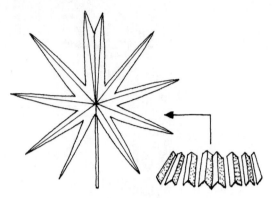

Reduplicate
外向折叠的

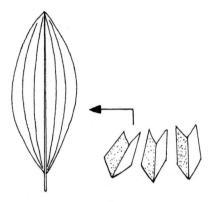

Induplicate
内向折叠的

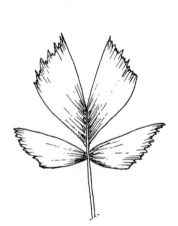

Bijugate
两对的

Trijugate
三对的

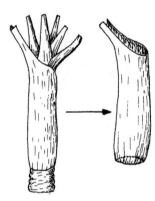

Crownshaft (*Roystonea* sp.)
冠轴（王棕属某植物）

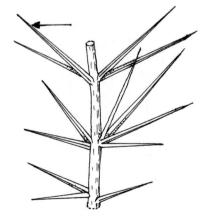

Acanthophyll (*Phoenix* sp.)
叶刺（刺葵属某植物）

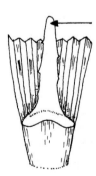

Hastula (*Copernicia* sp.)
小戟突（蜡棕属某植物）

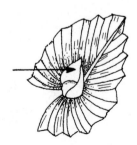

Cupular hastula (*Livistona* sp.)
杯状小戟突（蒲葵属某植物）

Structure of Palms 3
棕榈科的结构 3

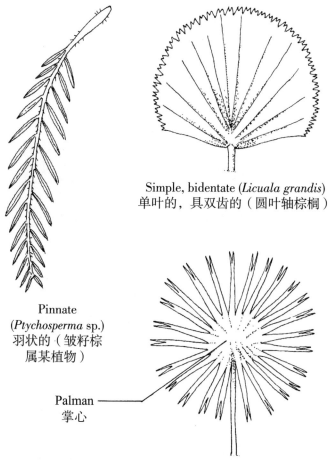

Pinnate
(*Ptychosperma* sp.)
羽状的（皱籽棕
属某植物）

Palman
掌心

Single folded segments with bidentate tips (*Chamaerops* sp.)
具两齿顶端的一次折叠的裂片（欧洲矮棕属某植物）

Simple, bidentate (*Licuala grandis*)
单叶的，具双齿的（圆叶轴棕榈）

Bipinnate with fish-tail
leaflets (*Caryota* sp.)
具鱼尾小叶的二回羽状的
（鱼尾葵属某植物）

Split centrally to base, segments
with several folds (*Chelyocarpus* sp.)
向中心裂至基部，裂片具数次折
叠（龟果棕属某植物）

Palmate (*Livistona* sp.)
掌状的（蒲葵属某植物）

Costa
中脉

Costapalmate (*Sabal* sp.)
具肋掌状的（菜棕属某植物）

Structure of Orchids 1
兰科的结构 1

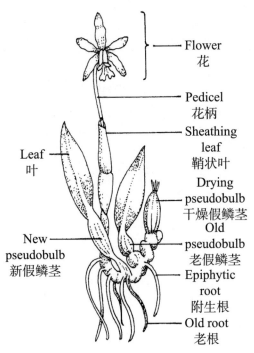

Flower
花

Pedicel
花柄

Sheathing leaf
鞘状叶

Drying pseudobulb
干燥假鳞茎

Old pseudobulb
老假鳞茎

Leaf
叶

New pseudobulb
新假鳞茎

Epiphytic root
附生根

Old root
老根

Parts of an epiphytic orchid (*Cattleya* sp.)
附生兰的部分（卡特兰属某植物）

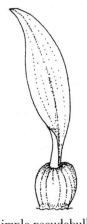

Simple pseudobulb bearing a single leaf
生有单叶的单假鳞茎

Compound pseudobulb
复合假鳞茎

Simple pseudobulb bearing 2 leaves
生有2叶的单假鳞茎

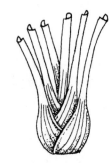

Compound pseudobulb enclosed in leaf bases
被叶基包住的复合假鳞茎

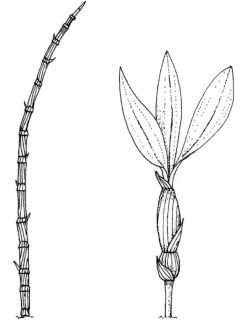

Cane-like stem
秆状茎

Fusiform compound pseudobulb
纺锤形复合假鳞茎

Ovoid compound pseudobulb
卵球形复合假鳞茎

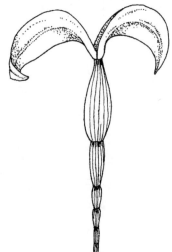

Clavate compound pseudobulb
棍棒形复合假鳞茎

Structure of Orchids 2
兰科的结构 2

THREE TYPES OF FLOWER ORIENTATION
花生长走向的三种类型

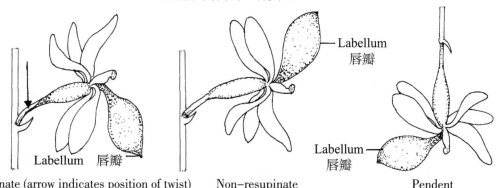

Resupinate (arrow indicates position of twist)
扭转的（箭头示扭转的位置）

Non-resupinate
不扭转的

Pendent
下垂的

FLOWER STRUCTURE 花的结构

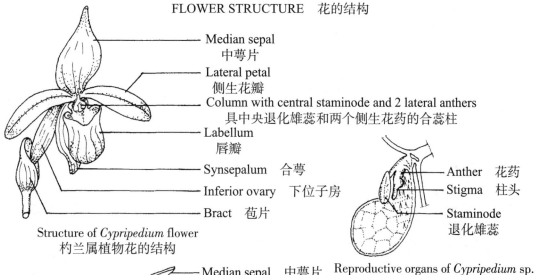

Median sepal
中萼片

Lateral petal
侧生花瓣

Column with central staminode and 2 lateral anthers
具中央退化雄蕊和两个侧生花药的合蕊柱

Labellum
唇瓣

Synsepalum 合萼

Inferior ovary 下位子房

Bract 苞片

Structure of *Cypripedium* flower
杓兰属植物花的结构

Anther 花药

Stigma 柱头

Staminode
退化雄蕊

Reproductive organs of *Cypripedium* sp.
杓兰属某植物的生殖器官

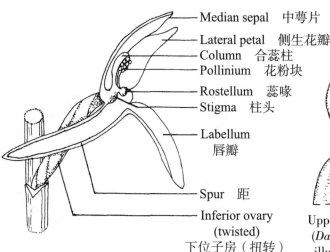

Median sepal 中萼片

Lateral petal 侧生花瓣

Column 合蕊柱

Pollinium 花粉块

Rostellum 蕊喙

Stigma 柱头

Labellum
唇瓣

Spur 距

Inferior ovary
(twisted)
下位子房（扭转）

L.S. of flower of *Dactylorhiza fuchsii*
斑点舌喙兰花的纵切

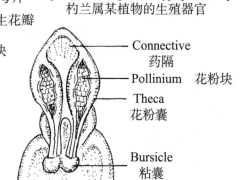

Connective
药隔

Pollinium 花粉块

Theca
花粉囊

Bursicle
粘囊

Upper portion of column showing stamen
(*Dactylorhiza fuchsii*) (see next page for
illustration of pollinium separated out)
合蕊柱上部示雄蕊（斑点舌喙兰）
（分出的花粉块图解，见下页）

Structure of Orchids 3
兰科的结构 3

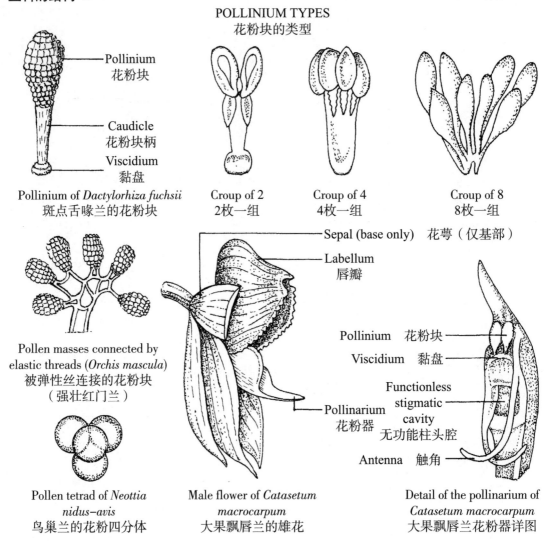

POLLINIUM TYPES
花粉块的类型

Pollinium
花粉块

Caudicle
花粉块柄

Viscidium
黏盘

Pollinium of *Dactylorhiza fuchsii*
斑点舌喙兰的花粉块

Croup of 2
2枚一组

Croup of 4
4枚一组

Croup of 8
8枚一组

Pollen masses connected by
elastic threads (*Orchis mascula*)
被弹性丝连接的花粉块
（强壮红门兰）

Sepal (base only) 花萼（仅基部）

Labellum
唇瓣

Pollinium 花粉块

Viscidium 黏盘

Functionless
stigmatic
cavity
无功能柱头腔

Pollinarium
花粉器

Antenna 触角

Pollen tetrad of *Neottia
nidus-avis*
鸟巢兰的花粉四分体

Male flower of *Catasetum
macrocarpum*
大果飘唇兰的雄花

Detail of the pollinarium of
Catasetum macrocarpum
大果飘唇兰花粉器详图

FLORAL DIAGRAMS OF THE MAJOR GROUPS
主要类群的花图式

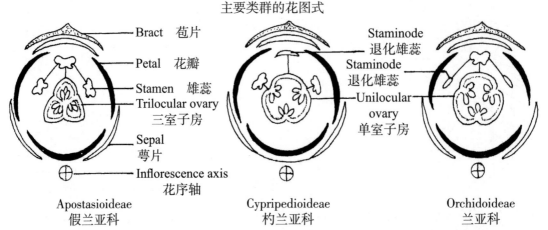

Bract 苞片

Petal 花瓣

Stamen 雄蕊

Trilocular ovary
三室子房

Sepal
萼片

Inflorescence axis
花序轴

Staminode
退化雄蕊

Staminode
退化雄蕊

Unilocular
ovary
单室子房

Apostasioideae
假兰亚科

Cypripedioideae
杓兰亚科

Orchidoideae
兰亚科

Structure of Orchids 4
兰科的结构 4

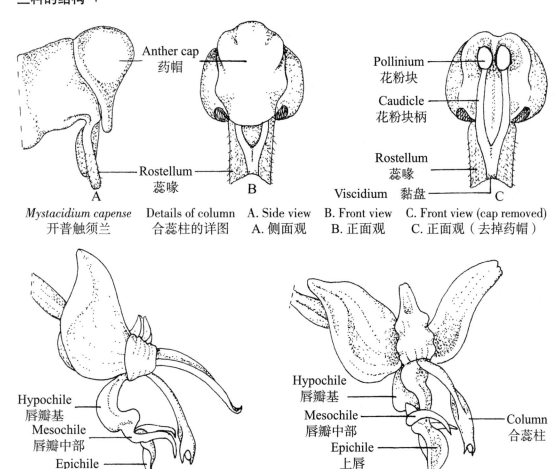

Anther cap 药帽	Pollinium 花粉块	
	Caudicle 花粉块柄	
Rostellum 蕊喙	Rostellum 蕊喙	
	Viscidium 黏盘	

Mystacidium capense 开普触须兰   Details of column 合蕊柱的详图   A. Side view A. 侧面观   B. Front view B. 正面观   C. Front view (cap removed) C. 正面观（去掉药帽）

Hypochile 唇瓣基
Mesochile 唇瓣中部
Epichile 上唇

Hypochile 唇瓣基
Mesochile 唇瓣中部
Epichile 上唇
Column 合蕊柱

Two species of *Stanhopea* showing details of column (side and front views)
奇唇兰属两种植物，示合蕊柱详图（侧面观和正面观）

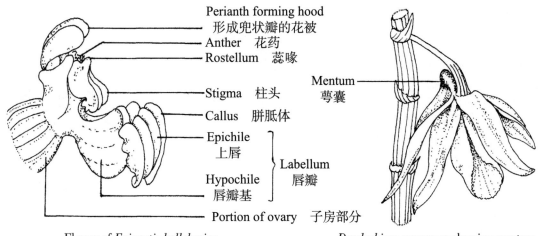

Perianth forming hood 形成兜状瓣的花被
Anther 花药
Rostellum 蕊喙
Stigma 柱头
Callus 胼胝体
Epichile 上唇
Hypochile 唇瓣基
} Labellum 唇瓣
Portion of ovary 子房部分

Mentum 萼囊

Flower of *Epipactis helleborine*
火烧兰的花

Dendrobium anosmum showing mentum
卓花石斛，示萼囊

12. Fruits
12. 果实

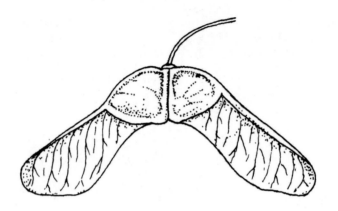

Vexillum Positions and Fruits of Leguminosae (Fabaceae)
旗瓣的位置和豆科（蝶形花科）的果实

VEXILLUM POSITIONS　旗瓣的位置

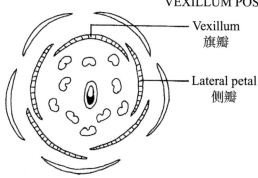

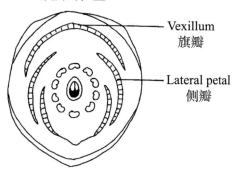

Imbricate–ascending in subfamily
Caesalpinioideae
苏木亚科的上向覆瓦状排列

Imbricate–descending (vexillary) in subfamily
Papilionoideae
蝶形花亚科的下向覆瓦状排列（蝶形卷叠式）

FRUITS 1
果实 1

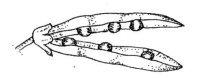

Legume (*Vicia* sp.)
荚果（野豌豆属某植物）

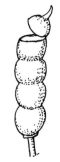

Lomentum (*Ornithopus* sp.)
节荚（鸟足豆属某植物）

4–winged pod of *Tetragonolobus purpureus*
四棱豆的4翅荚果

Reticulate pod of *Arachis hypogaea*
落花生的网状荚果

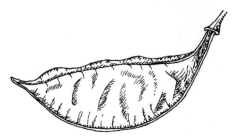

Inflated many–seeded pod with papery
walls of *Colutea arborescens*
鱼鳔槐的具纸质壁、膨大的、含多
数种子的荚果

Spirally coiled pod of
Medicago sativa
紫苜蓿螺旋卷曲的荚果

1– or 2–seeded pod with beak of
Trigonella caerulea
卢豆具喙、含1或2枚种子的荚果

Fruits 2
果实 2

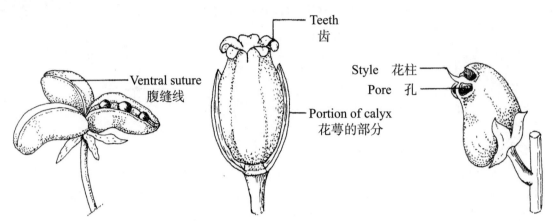

Teeth
齿

Style　花柱
Pore　孔

Ventral suture
腹缝线

Portion of calyx
花萼的部分

Follicle (*Paeonia* sp.)　　Capsule with apical teeth (*Silene* sp.)　Poricidal capsule (*Antirrhinum* sp.)
菁葵果（芍药属某植物）顶端具齿的蒴果（蝇子草属某植物）孔裂的蒴果（金鱼草属某植物）

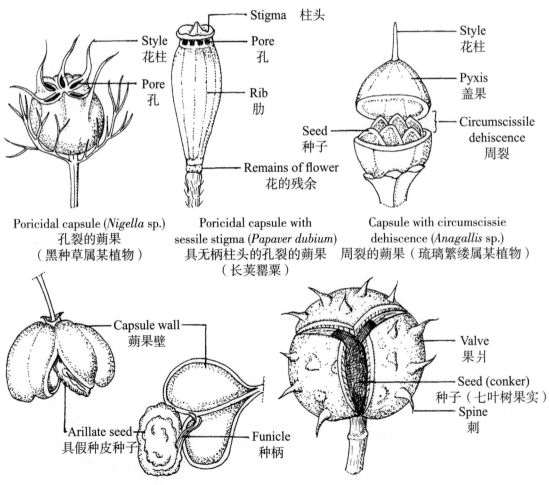

Style
花柱

Pore
孔

Stigma　柱头
Pore
孔
Rib
肋
Seed
种子
Remains of flower
花的残余

Style
花柱
Pyxis
盖果
Seed
种子
Circumscissile
dehiscence
周裂

Poricidal capsule (*Nigella* sp.)
孔裂的蒴果
（黑种草属某植物）

Poricidal capsule with
sessile stigma (*Papaver dubium*)
具无柄柱头的孔裂的蒴果
（长荚罂粟）

Capsule with circumscissie
dehiscence (*Anagallis* sp.)
周裂的蒴果（琉璃繁缕属某植物）

Capsule wall
蒴果壁

Valve
果爿

Seed (conker)
种子（七叶树果实）

Spine
刺

Arillate seed
具假种皮种子

Funicle
种柄

4–lobed capsule dehiscing longitudinally, whole and in
section (*Euonymus europaeus*)
纵向开裂的4浅裂蒴果，整体和剖面（欧洲卫矛）

Valvate capsule (*Aesculus hippocastanum*)
瓣裂的蒴果（欧洲七叶树）

Fruits 3
果实 3

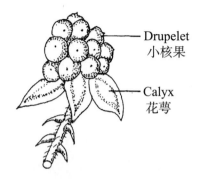

Etaerio of drupelets (*Rubus* sp.)
具小核果的聚心皮果
（悬钩子属某植物）

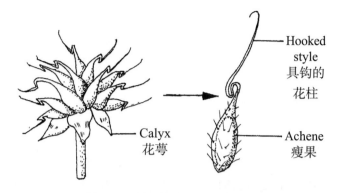

Etaerio of hooked achenes (*Geum* sp.)
具钩瘦果的聚心皮果（路边青属某植物）

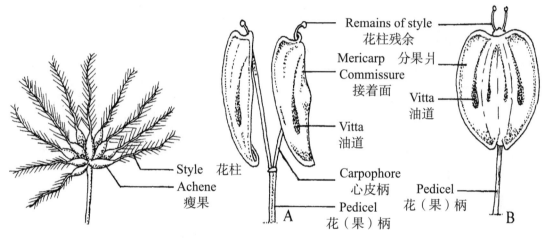

Etaerio of feathered achenes (*Clematis vitalba*)
羽毛状瘦果的聚心皮果（葡匐叶铁线莲）

Heracleum sphondylium A. The 2 mericarps (one twisted round) pendent on the divided carpophore
B. The whole schizocarp before dehiscence
欧白芷 A. 2枚分果爿（一面扭曲的）悬挂在分开的心皮柄上 B. 开裂前的整个分果

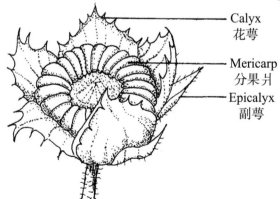

Schizocarp of *Malva sylvestris*
欧锦葵的分果

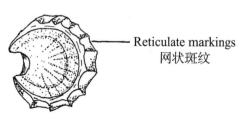

Mericarp of *Malva sylvestris*, a wedge-shaped nutlet
欧锦葵的分果爿，一个楔形的小坚果

Fruits 4
果实 4

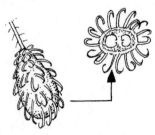

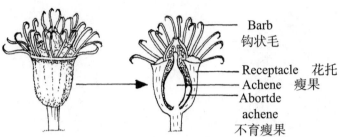

Barb
钩状毛

Receptacle　花托
Achene　瘦果
Abortde
achene
不育瘦果

Hooked pericarp, whole and in section
(*Circaea lutetiana*)
具钩的果皮，整体和剖面（水珠草）

Barbs developed on the receptacle, whole and in section
(*Agrimonia eupatoria*)
果托上发育的钩状毛，整体和剖面（欧洲龙芽草）

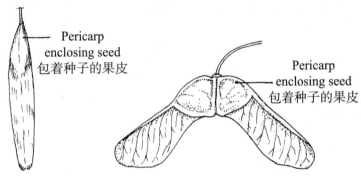

Pericarp
enclosing seed
包着种子的果皮

Pericarp
enclosing seed
包着种子的果皮

Involucre of hooked bracts
surrounding infructescence
(*Arctium minus*)
围绕果序周围的、由具钩苞
片构成的总苞（小牛蒡）

Samara (*Fraxinus excelsior*)
翅果（欧洲白蜡）

Double samara (*Acer* sp.)
双悬翅果（槭属某植物）

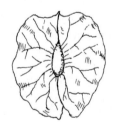

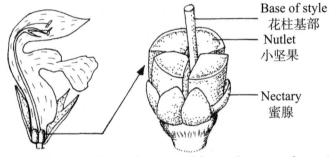

Base of style
花柱基部
Nutlet
小坚果

Nectary
蜜腺

Winged pericarp (*Ulmus* sp.)
具翅的果皮（榆属某植物）

Nutlets of *Lamium album* with section of flower showing nutlet position
短柄野芝麻的小坚果，附花的剖面图，示小坚果位置

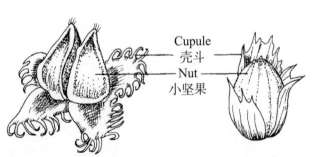

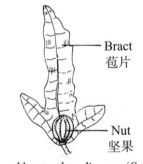

Cupule
壳斗
Nut
小坚果

Bract
苞片

Nut
坚果

Triangular nuts (*Fagus sylvatica*)
三棱形坚果（欧洲山毛榉）

Nuts (*Corylus avellana*)
坚果（欧洲榛子）

Winged bract subtending nut (*Carpinus betulus*)
包着坚果的具翅苞片（欧洲鹅耳枥）

Fruits 5
果实 5

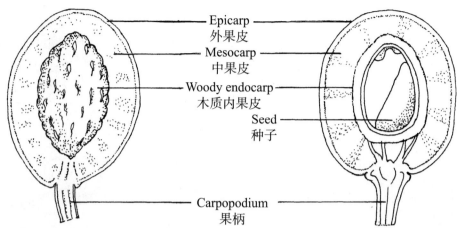

Epicarp
外果皮

Mesocarp
中果皮

Woody endocarp
木质内果皮

Seed
种子

Carpopodium
果柄

Portion of drupe cut away to expose outer surface of
stone (*Prunus* sp.)
切开核果部分，露出核的外表面（李属某植物）

L.S. of drupe (*Prunus* sp.)
核果的纵切（李属某植物）

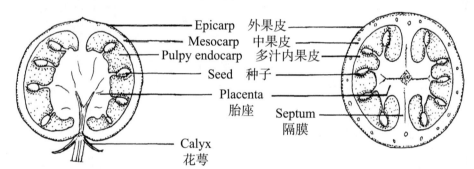

Epicarp 外果皮
Mesocarp 中果皮
Pulpy endocarp 多汁内果皮
Seed 种子
Placenta 胎座
Septum 隔膜
Calyx
花萼

L.S. of berry (*Lycopersicon* sp.)
浆果的纵切（番茄属某植物）

T.S. of berry (*Lycopersicon* sp.)
浆果的横切（番茄属某植物）

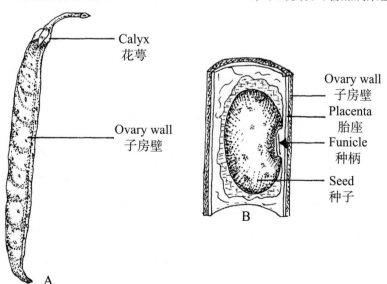

Calyx
花萼

Ovary wall
子房壁

Ovary wall
子房壁

Placenta
胎座

Funicle
种柄

Seed
种子

A

B

Phaseolus coccineus　A. Whole legume　B. Portion of legume opened up to show attachment of seed
荷包豆　A. 整个荚果　B. 剖开的荚果部分，示种子附着情况

Fruits 6
果实 6

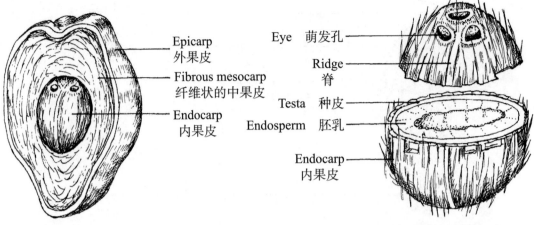

Epicarp
外果皮

Fibrous mesocarp
纤维状的中果皮

Endocarp
内果皮

Eye　萌发孔

Ridge
脊

Testa　种皮

Endosperm　胚乳

Endocarp
内果皮

Fibrous drupe of *Cocos nucifera* with part of fruit wall
removed to expose endocarp
部分果壁被剥去露出内果皮的椰子纤维质核果

T.S. of endocarp exposing seed
(fruit inverted for clarity)
露出种子的内果皮横切
（为更清楚将果实倒置）

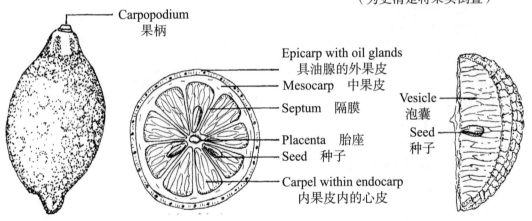

Carpopodium
果柄

Epicarp with oil glands
具油腺的外果皮

Mesocarp　中果皮

Septum　隔膜

Placenta　胎座

Seed　种子

Carpel within endocarp
内果皮内的心皮

Vesicle
泡囊

Seed
种子

Hesperidium of *Citrus limon*
柠檬的柑果

T.S. of fruit
果实的横切

Single carpel removed from endocarp
从内果皮剥出的单心皮

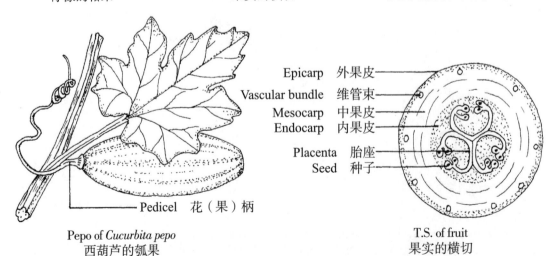

Pedicel　花（果）柄

Epicarp　外果皮

Vascular bundle　维管束

Mesocarp　中果皮

Endocarp　内果皮

Placenta　胎座

Seed　种子

Pepo of *Cucurbita pepo*
西葫芦的瓠果

T.S. of fruit
果实的横切

Fruits 7
果实 7

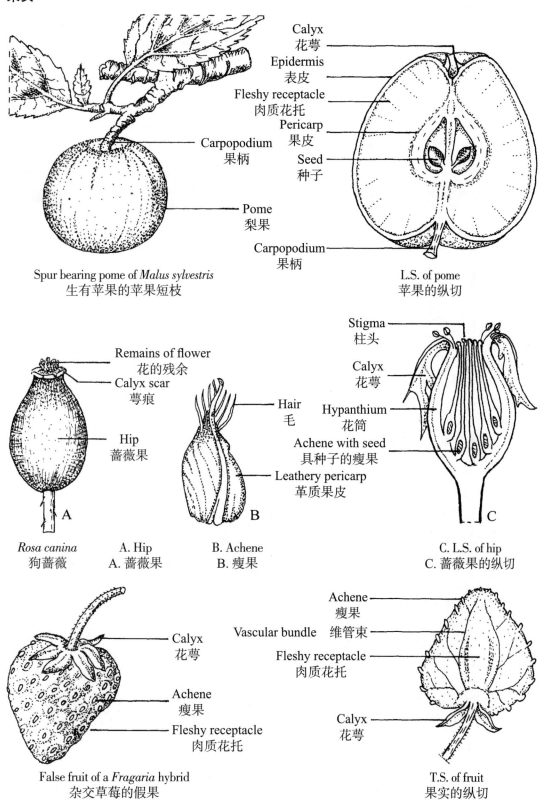

Calyx
花萼

Epidermis
表皮

Fleshy receptacle
肉质花托

Pericarp
果皮

Seed
种子

Carpopodium
果柄

Pome
梨果

Carpopodium
果柄

Spur bearing pome of *Malus sylvestris*
生有苹果的苹果短枝

L.S. of pome
苹果的纵切

Remains of flower
花的残余

Calyx scar
萼痕

Hip
蔷薇果

Hair
毛

Leathery pericarp
革质果皮

Stigma
柱头

Calyx
花萼

Hypanthium
花筒

Achene with seed
具种子的瘦果

Rosa canina
狗蔷薇

A. Hip
A. 蔷薇果

B. Achene
B. 瘦果

C. L.S. of hip
C. 蔷薇果的纵切

Calyx
花萼

Achene
瘦果

Fleshy receptacle
肉质花托

Achene
瘦果

Vascular bundle
维管束

Fleshy receptacle
肉质花托

Calyx
花萼

False fruit of a *Fragaria* hybrid
杂交草莓的假果

T.S. of fruit
果实的纵切

Fruits 8
果实 8

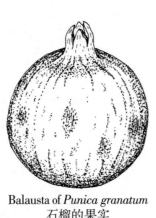

Balausta of *Punica granatum*
石榴的果实

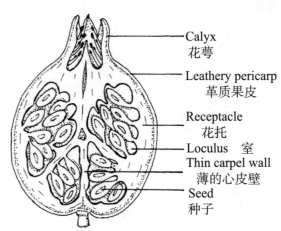

Calyx
花萼

Leathery pericarp
革质果皮

Receptacle
花托

Loculus 室

Thin carpel wall
薄的心皮壁

Seed
种子

L.S. of balausta
石榴果实的纵切

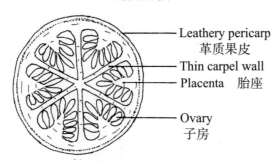

Leathery pericarp
革质果皮

Thin carpel wall
薄的心皮壁

Placenta 胎座

Ovary
子房

T.S. of balausta
石榴果实的横切

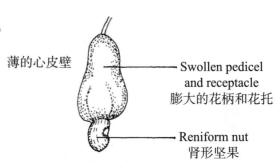

Swollen pedicel
and receptacle
膨大的花柄和花托

Reniform nut
肾形坚果

Fruit of *Anacardium occidentale*
腰果的果实

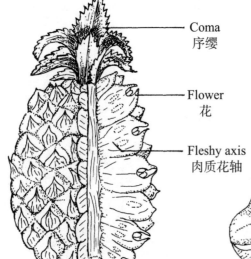

Coma
序缨

Flower
花

Fleshy axis
肉质花轴

Syncarp (coenocarpium) of *Ananas comosus*
菠萝的聚花果（聚合果）

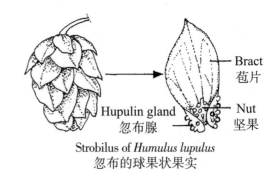

Bract
苞片

Hupulin gland
忽布腺

Nut
坚果

Strobilus of *Humulus lupulus*
忽布的球果状果实

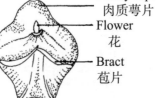

Fleshy sepal
肉质萼片

Flower
花

Bract
苞片

Individual flower
单花

Syncarp (sorosis) of *Morus nigra*
黑桑的聚花果（椹果）

Fruits 9　Syconium
果实 9　隐头果

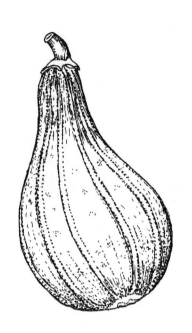

Syconium of *Ficus carica*
无花果的隐头果

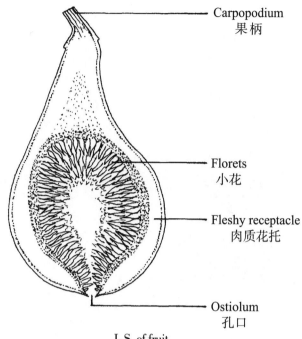

Carpopodium
果柄

Florets
小花

Fleshy receptacle
肉质花托

Ostiolum
孔口

L.S. of fruit
果实的纵切

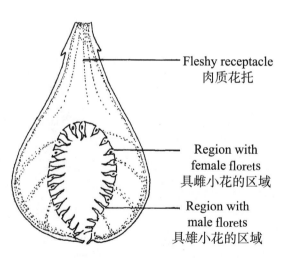

Fleshy receptacle
肉质花托

Region with
female florets
具雌小花的区域

Region with
male florets
具雄小花的区域

L.S. of fruit–diagram showing
distribution of male and female florets
果实纵切——揭示雄小花和雌小
花分布的示意图

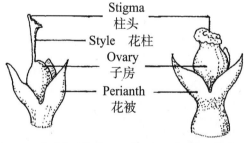

Stigma
柱头
Style　花柱
Ovary
子房
Perianth
花被

Long–styled female floret
具长花柱的雌小花

Short–styled female floret
(becomes a gall)
具短花柱的雌小花
（变为虫瘿）

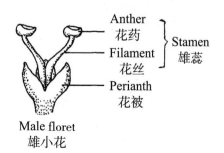

Anther
花药
Filament
花丝
} Stamen
雄蕊
Perianth
花被

Male floret
雄小花

The 3 types of florets
小花的三种类型

Fruits 10   Cruciferae (Brassicaceae) 1
果实 10   十字花科 1

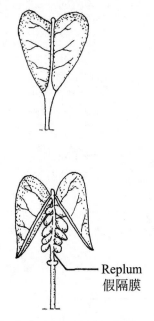

Silicula before and after dehiscence
(*Capsella bursa-pastoris*)
开裂前和开裂后的短角果
（荠菜）

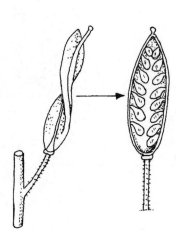

Twisted siliqua at early
dehiscence (*Draba incana*)
早期开裂时扭曲的长角果
（灰白葶苈）

Replum
假隔膜

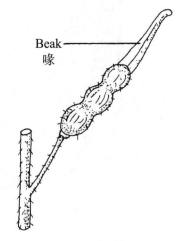

Beak
喙

Siliqua with long beak
(*Sinapis alba*)
具长喙的长角果
（白芥）

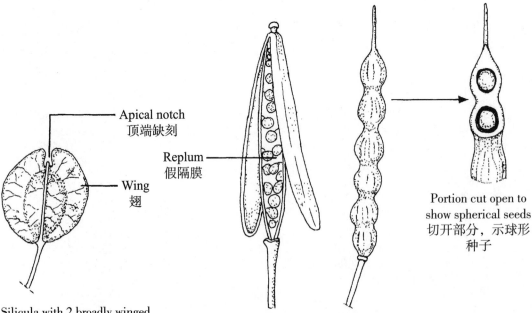

Apical notch
顶端缺刻

Replum
假隔膜

Wing
翅

Portion cut open to
show spherical seeds
切开部分，示球形
种子

Silicula with 2 broadly winged
valves (*Thlaspi arvense*)
具2宽翅果爿的短角果
（菥蓂）

Flattened siliqua
(*Erysimum cheiri*)
扁平长角果（红花糖芥）

Lomentum-like fruit
(*Raphanus raphanistrum*)
节荚般的果实（野萝卜）

Fruits 11   Cruciferae (Brassicaceae) 2
果实 11   十字花科 2

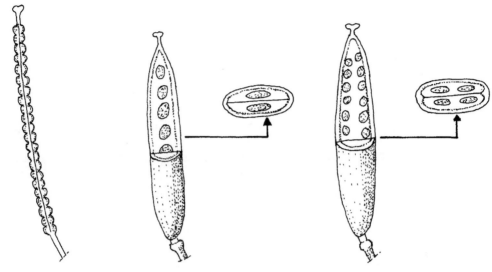

Torulose (*Erysimum repandum*)
念珠状的（粗梗糖芥）

Uniseriate seed arrangement
单列种子的排列

Biseriate seed arrangement
两列种子的排列

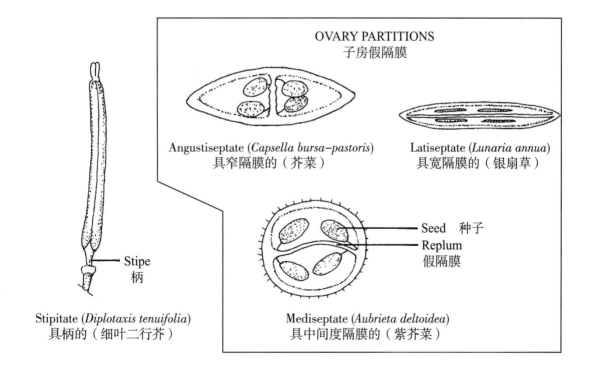

OVARY PARTITIONS
子房假隔膜

Angustiseptate (*Capsella bursa-pastoris*)
具窄隔膜的（芥菜）

Latiseptate (*Lunaria annua*)
具宽隔膜的（银扇草）

Stipe
柄

Seed　种子
Replum
假隔膜

Stipitate (*Diplotaxis tenuifolia*)
具柄的（细叶二行芥）

Mediseptate (*Aubrieta deltoidea*)
具中间度隔膜的（紫芥菜）

13. Conifers and Conifer Allies
13. 球果类与拟球果类

Conifers 1 *Pinus sylvestris*
球果类 1 长白松

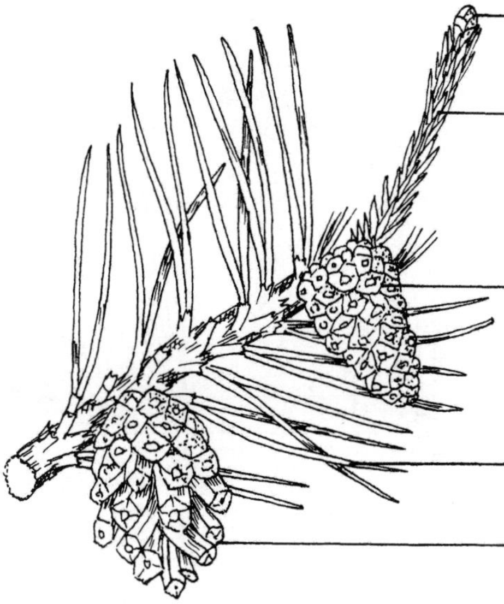

Young female cone
幼小雌球花

Young dwarf shoots
幼小短枝

Green 2nd year cone
绿色的第二年球果

Paired needles
成对的针叶

Woody 3rd year seed-shedding cone
第三年种子脱落的木质球果

Female cones
雌球花

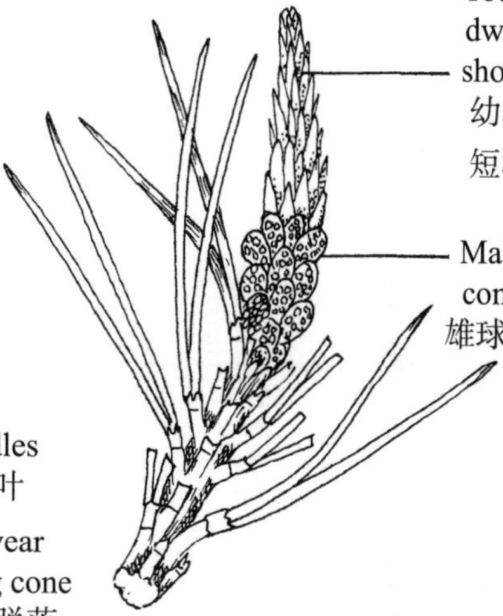

Young dwarf shoots
幼小短枝

Male cone
雄球花

Male cones
雄球花

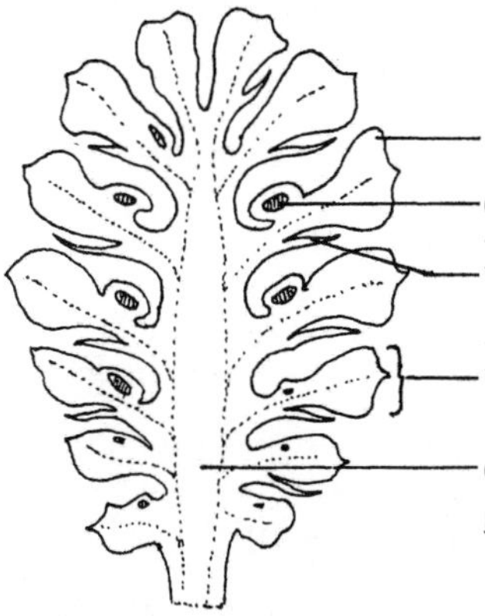

Ovuliferous scale
珠鳞

Ovule 胚珠

Bract scale
苞鳞

Megasporophyll
大孢子叶

Cone axis
雌球花轴

L.S. of female cone
雌球花纵切

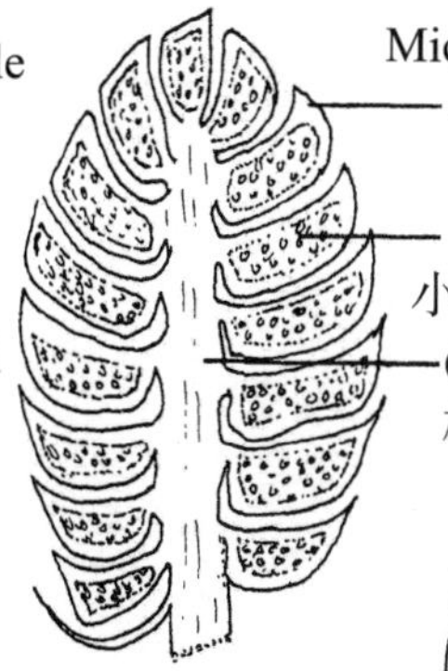

Microsporophyll
小孢子叶

Microsporangium (with pollen grains)
小孢子囊（具有花粉粒）

Cone axis
雄球花轴

L.S. of male cones
雄球花纵切

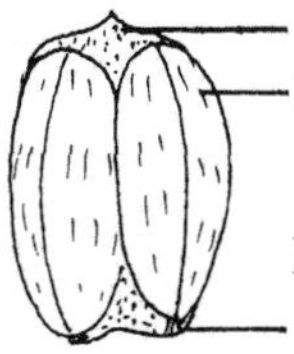

Microsporophyll
小孢子叶

Microsporangium
小孢子囊

Point of attachment
着生点

Underside of Microsporophyll
小孢子叶的背面

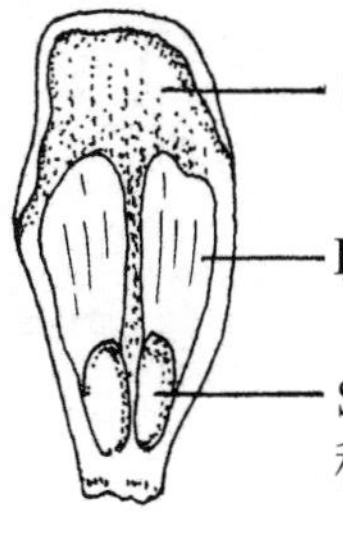

Seed(cone) scale
种（果）鳞

Papery wing
纸质翅

Seed
种子

Mature seeds resting on a cone scale
留在果鳞上的成熟种子

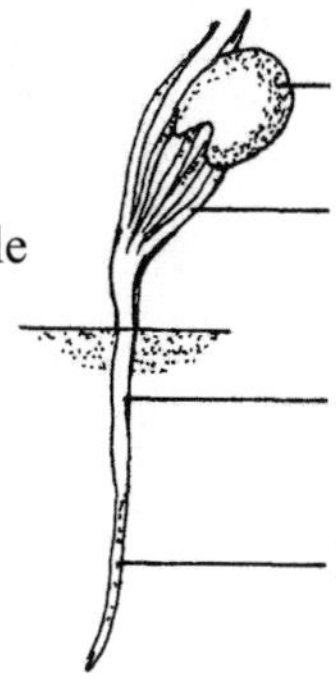

Testa
种皮

Cotyledon
子叶

Hypocotyl
下胚轴

Radicle
胚根

Seedling
幼苗

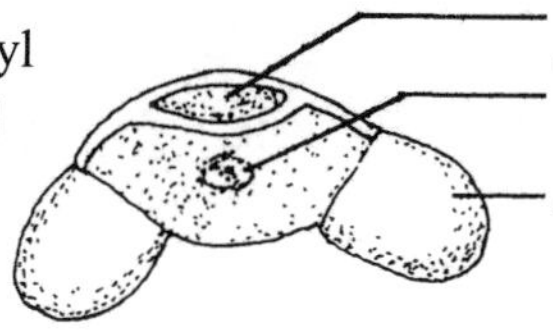

Prothallial cells
原叶细胞

Vegetative nucleus
营养核

Air sac in exine
外壁上的气囊

Pollen grain
花粉粒

Conifers 2
球果类 2

PINE NEEDLE ARRANGEMENT　松针的排列

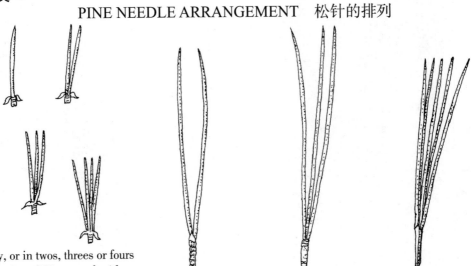

Singly, or in twos, threes or fours
Varieties of *Pinus cembroides*
单针，或两针一束，三针或
四针一束
（墨西哥石松的变种）

In twos (*Pinus nigra*)　In threes (*Pinus radiata*)　In fives (*Pinus peuce*)
两针一束（黑松）　　三针一束（辐射松）　五针一束（马其顿松）

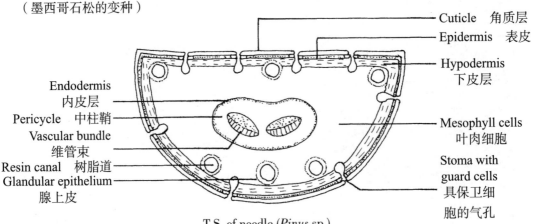

Cuticle　角质层

Epidermis　表皮

Hypodermis
下皮层

Endodermis
内皮层
Pericycle　中柱鞘
Vascular bundle
维管束
Resin canal　树脂道
Glandular epithelium
腺上皮

Mesophyll cells
叶肉细胞

Stoma with
guard cells
具保卫细
胞的气孔

T.S. of needle (*Pinus* sp.)
松针的横切（松属某植物）

Leaves pectinately arranged
(*Metasequoia glyptostroboides*)
篦齿状排列的叶（水杉）

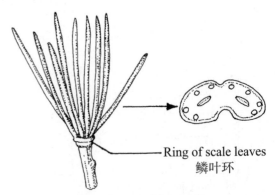

Ring of scale leaves
鳞叶环

Leaves arranged in whorls
(*Sciadopitys verticillata*)
轮状排列的叶（金松）

Conifers 3
球果类 3

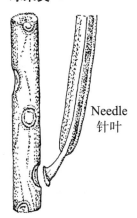

Needle
针叶

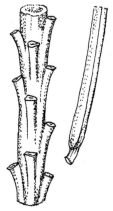

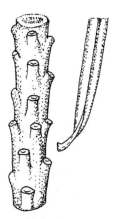

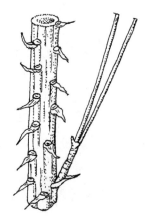

Concave scars
(*Abies* sp.)
凹陷的叶痕
（冷杉属某植物）

Woody pulvinus
(*Picea* sp.)
木质的叶枕（云
杉属某植物）

Raised scars
(*Pseudotsuga* sp.)
凸起的叶痕（黄
杉属某植物）

Spur shoot with
facsicles (*Pinus* sp.)
具呈束针叶的短枝
（松属某植物）

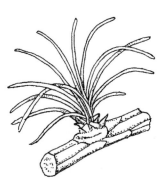

Underside
of leaf
叶的背面

Clustered needles on
short shoot (*Larix* sp.)
短枝上的簇生针叶
（落叶松属某植物）

Spirally arranged broad
leaves (*Araucaria araucana*)
螺旋状排列的阔叶（智利
南洋杉）

Scale leaves with
ridge (*Thuja* sp.)
具脊的鳞叶（崖
柏属某植物）

White bands of stomata
(*Fokienia hodginsii*)
白色气孔带
（福建柏）

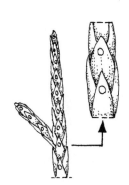

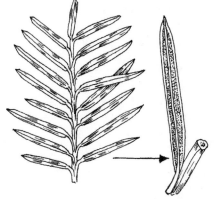

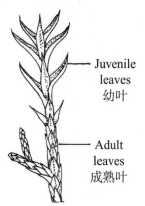

Juvenile
leaves
幼叶

Adult
leaves
成熟叶

Resin gland on scale
leaves (*Chamaecyparis* sp.)
鳞叶上的树脂腺（扁柏
属某植物）

Upper and lower surface of *Taxus* leaves,
spirally arranged but appearing as 2-ranked
红豆杉属植物叶的上、下表面，螺旋状
排列但呈2列

Two forms of leaves
(*Juniperus* sp.)
叶的两种类型
（刺柏属某植物）

Conifers 4
球果类 4

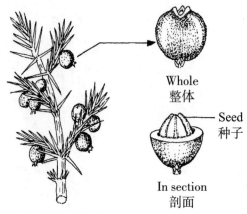

Whole
整体

Seed
种子

In section
剖面

Galbuli–the fruits of *Juniperus communis*
球果—欧洲刺柏的果实

Terminal cone (*Sequoiadendron giganteum*)
顶生球果（巨杉）

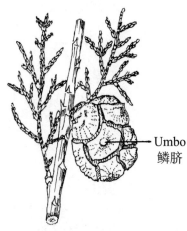

Umbo
鳞脐

Ovoid cone on short stalk
(*Cupressus sempervirens*)
短柄上的卵球形球果
（意大利柏）

Upright cone scales
(*Larix decidua*)
直立果鳞（欧洲落叶松）

Bracts between cone scales
(*Pseudotsuga menziesii*)
果鳞间的苞片（花旗松）

Reflexed cone scales
(*Larix kaempferi*)
反折的果鳞（日本落叶松）

Ripe cone, globose when young
(*Chamaecyparis lawsoniana*)
成熟球果，幼时球形（美国扁柏）

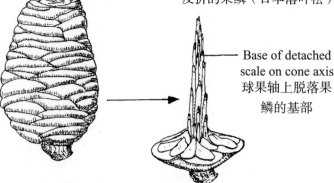

Base of detached
scale on cone axis
球果轴上脱落果
鳞的基部

Barrel–shaped cone, before and after breaking up (*Cedrus libani*)
桶形球果，开裂前和开裂后（黎巴嫩雪松）

Thin flexible cone with
reflexed scales (*Thuja plicata*)
生有反折果鳞的细而柔韧
的球果（北美香柏）

Conifer Allies 1 *Ginkgo biloba*
拟球果类 1 银杏

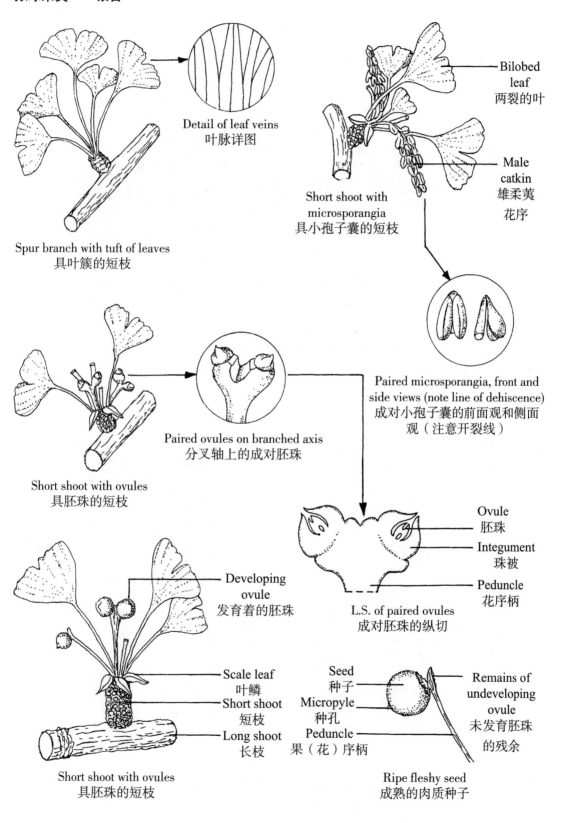

Detail of leaf veins
叶脉详图

Spur branch with tuft of leaves
具叶簇的短枝

Short shoot with
microsporangia
具小孢子囊的短枝

Bilobed
leaf
两裂的叶

Male
catkin
雄柔荑
花序

Paired microsporangia, front and
side views (note line of dehiscence)
成对小孢子囊的前面观和侧面
观（注意开裂线）

Paired ovules on branched axis
分叉轴上的成对胚珠

Short shoot with ovules
具胚珠的短枝

Developing
ovule
发育着的胚珠

Ovule
胚珠

Integument
珠被

Peduncle
花序柄

L.S. of paired ovules
成对胚珠的纵切

Scale leaf
叶鳞

Short shoot
短枝

Long shoot
长枝

Short shoot with ovules
具胚珠的短枝

Seed
种子

Micropyle
种孔

Peduncle
果（花）序柄

Remains of
undeveloping
ovule
未发育胚珠
的残余

Ripe fleshy seed
成熟的肉质种子

Conifer Allies 2　Cycad
拟球果类 2　铁树

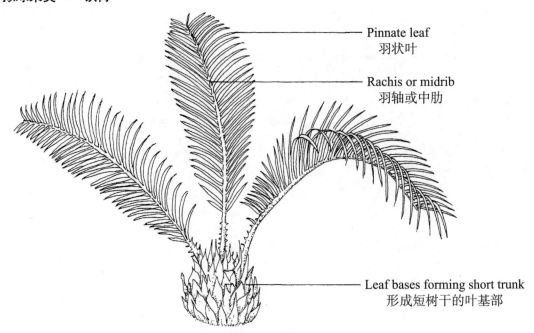

Pinnate leaf
羽状叶

Rachis or midrib
羽轴或中肋

Leaf bases forming short trunk
形成短树干的叶基部

Whole plant before maturity (only 3 leaves shown)
成熟前的整体植株（仅示3枚叶）

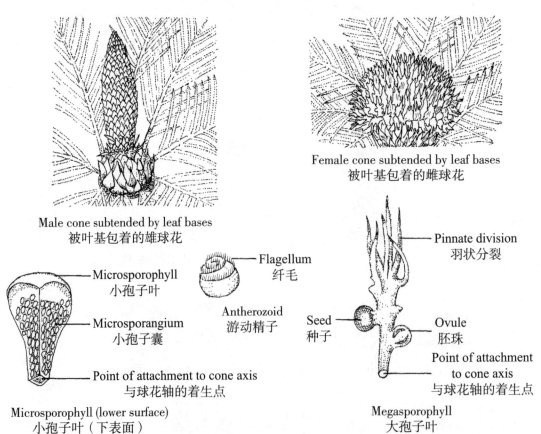

Female cone subtended by leaf bases
被叶基包着的雌球花

Male cone subtended by leaf bases
被叶基包着的雄球花

Microsporophyll
小孢子叶

Microsporangium
小孢子囊

Point of attachment to cone axis
与球花轴的着生点

Flagellum
纤毛

Antherozoid
游动精子

Pinnate division
羽状分裂

Seed
种子

Ovule
胚珠

Point of attachment
to cone axis
与球花轴的着生点

Microsporophyll (lower surface)
小孢子叶（下表面）

Megasporophyll
大孢子叶

Conifer Allies 3
拟球果类 3

Node with scale leaves
具鳞叶的节

Ribbed stem
具纵棱的茎

Ephedra sp. Female shoot
麻黄属某植物的雌枝

Female shoot with 2 ovules
具2枚胚珠的雌枝

Fruit of *E. distachya* var. *helvetica*
瑞士双穗麻黄的果实

Stamen
雄蕊

Perianth
花被

Male flower
雄花

Integument tip
珠被顶端

Perianth
花被

Branchlet
小枝

Branchlet with terminal female flower
具顶生雌花的小枝

Stamen
雄蕊

Perianth
花被

Bract
苞片

Male shoot
雄枝

Male strobilus
雄球花

Stamen 雄蕊

Perianth 花被

Hair 毛

Male flower
雄花

Micropylar tube
珠孔管

Ovule
胚珠

Female flower
雌花

Collar
花座

Axis
花轴

Portion of male strobilus with numerous flowers
具多数花的雄球花部分

Female strobilus
雌球花

Female flower
雌花

Axis
花轴

Portion of female strobilus
雌球花部分

Gnetum gnemon Shoot with male inflorescences
直立买麻藤具雄花序的枝

Microsporangiate shoot
具小孢子囊的枝

Waxy leaf
蜡质叶

Stem 茎

Taproot
直根

Welwitschia mirabilis Male plant
百岁兰的雄株

14. Ferns and Fern Allies
14. 蕨类和拟蕨类

Ferns 1 *Dryopteris*
蕨类 1 鳞毛蕨

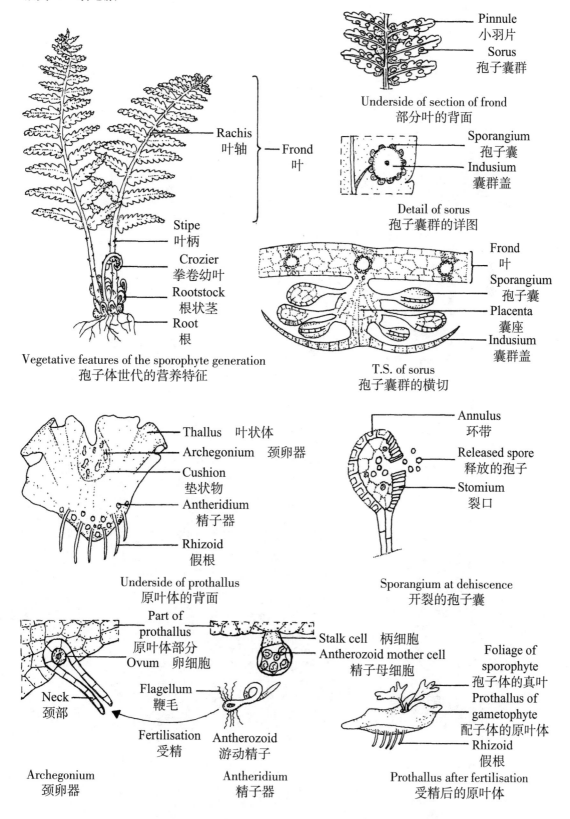

Pinnule
小羽片

Sorus
孢子囊群

Underside of section of frond
部分叶的背面

Sporangium
孢子囊

Indusium
囊群盖

Detail of sorus
孢子囊群的详图

Rachis
叶轴

Frond
叶

Frond
叶

Sporangium
孢子囊

Placenta
囊座

Indusium
囊群盖

T.S. of sorus
孢子囊群的横切

Stipe
叶柄

Crozier
拳卷幼叶

Rootstock
根状茎

Root
根

Vegetative features of the sporophyte generation
孢子体世代的营养特征

Thallus 叶状体

Archegonium 颈卵器

Cushion
垫状物

Antheridium
精子器

Rhizoid
假根

Underside of prothallus
原叶体的背面

Annulus
环带

Released spore
释放的孢子

Stomium
裂口

Sporangium at dehiscence
开裂的孢子囊

Part of
prothallus
原叶体部分

Ovum 卵细胞

Stalk cell 柄细胞

Antherozoid mother cell
精子母细胞

Neck
颈部

Flagellum
鞭毛

Fertilisation
受精

Antherozoid
游动精子

Foliage of
sporophyte
孢子体的真叶

Prothallus of
gametophyte
配子体的原叶体

Rhizoid
假根

Archegonium
颈卵器

Antheridium
精子器

Prothallus after fertilisation
受精后的原叶体

Ferns 2
蕨类 2

LEAF ARCHITECTURE
叶的结构

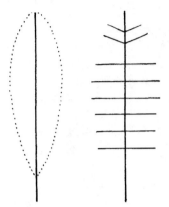  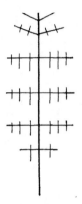 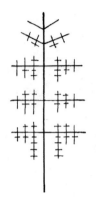 

Segment
羽片

Entire
(*Phyllitis* sp.)
全缘的（对开
蕨属某植物）

Pinnate
(*Oreopteris* sp.)
羽状的（假鳞毛
蕨属某植物）

Bipinnate
(*Dryopteris* sp.)
二回羽状的（鳞
毛蕨属某植物）

Tripinnate
(*Cryptogramma* sp.)
三回羽状的（珠
蕨属某植物）

Tripinnate with fan-shaped
segments (*Adiantum* sp.)
具扇形羽片的三回羽状
的（铁线蕨属某植物）

Entire
全缘的

Pinnate
羽状的

Pinnatifid
羽状半裂的

Bipinnate
二回羽状的

Bipinnatifid
二回羽状半裂的

 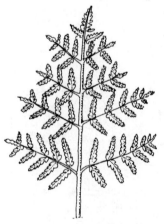 

Pinnulet
末回裂片
Sorus
孢子囊群

Tripinnate
三回羽状的

Tripinnatifid
三回羽状半裂的

Single pinnule
单枚小羽片

Ferns 3
蕨类 3

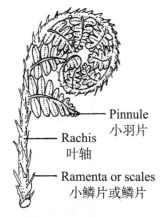

Pinnule
小羽片

Rachis
叶轴

Ramenta or scales
小鳞片或鳞片

Circinate vernation of
crozier (*Dryopteris* sp.)
蕨类的拳卷幼叶的拳卷卷
叠式（鳞毛蕨属某植物）

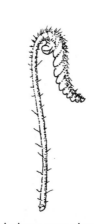

Non-circinate vernation of crozi
(*Pellaea falcata*)
蕨类的拳卷幼叶的非拳卷卷叠式
（镰叶旱蕨）

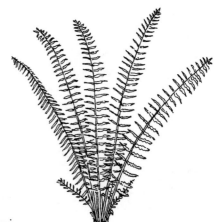

A 'shuttlecock' fern
"羽毛球形" 蕨

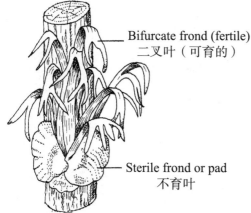

Bifurcate frond (fertile)
二叉叶（可育的）

Sterile frond or pad
不育叶

Epiphytic fern (*Platycerium* sp.)
附生蕨（鹿角蕨属某植物）

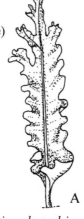

Phyllitis scolopendrium
对开蕨

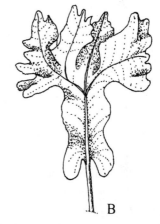

A

B

A. Crispate form
A. 皱波型

B. Crested form
B. 鸡冠型

Apospory in *Asplenium
bulbiferum* Plantlet
growing on pinna
铁角蕨的无孢子生殖，
羽片上生长的小植株

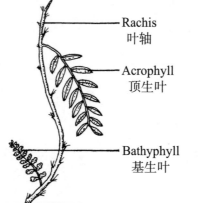

Rachis
叶轴

Acrophyll
顶生叶

Bathyphyll
基生叶

Acrophyll and bathyphyll of a climbing fern
攀缘蕨类的顶生叶和基生叶

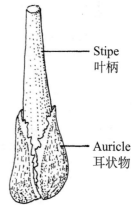

Stipe
叶柄

Auricle
耳状物

Auricles on *Angiopteris evecta*
观音座莲蕨的耳状物

Ferns 4
蕨类 4

STIPE JOINTS TO RHIZOMES
叶柄连到根状茎上

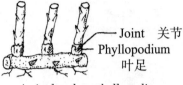

Joint　关节
Phyllopodium
叶足

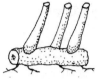

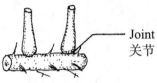

Joint
关节

Articulated on phyllopodium
叶足上具关节的

Direct connection
直接连接

Articulated
有关节的

VENATION
叶脉

Free
离生的

Forked
叉状的

Forked and free
叉状的和离生的

Pinnate
羽状的

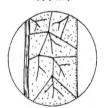

Anastomosing and free
网结的和离生的

Anastomosing
网结的

Anastomosing with veinlets
具小脉的网结的

Anastomosing with
branched veinlets
具分支小脉的网结的

PINNA ATTACHMENT
羽状着生方式

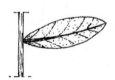

Attached by vein and base
以叶脉和叶基着生的

Decurrent
下延的

Surcurrent
上延的

Attached by vein only
仅以叶脉着生的

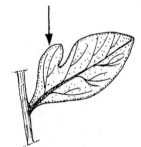

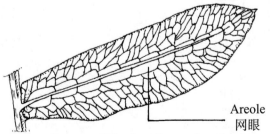

Areole
网眼

Pinna with acroscopic basal lobe ('thumb')
具上侧基生裂片（"拇指形"）的羽片

Enlarged pinna of *Woodwardia areolata* showing venation
网脉狗脊蕨的放大羽片，示脉序

Ferns 5
蕨类 5

INDUSIA　囊群盖

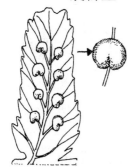

Reniform (*Dryopteris* sp.)
肾形的（鳞毛蕨属某植物）

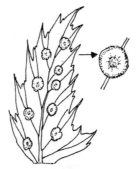

Peltate (*Polystichum* sp.)
盾形的（耳蕨属某植物）

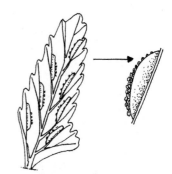

Elongate (*Asplenium* sp.)
长形的（铁角蕨属某植物）

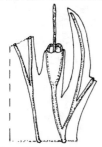

Cup–like or trumpet-shaped (*Trichomanes* sp.)
杯状或喇叭状的（单叶假脉蕨属某植物）

Valvate (*Hymenophyllum* sp.)
瓣裂的（膜蕨属某植物）

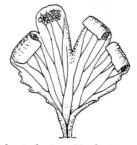

False indusium (*Asplenium* sp.)
假囊群盖（铁角蕨属某植物）

SORUS ARRANGEMENT　孢子囊群的排列

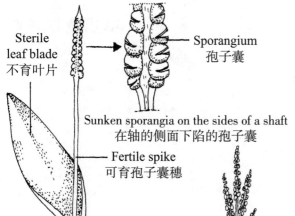

Sterile leaf blade
不育叶片

Sporangium
孢子囊

Sunken sporangia on the sides of a shaft
在轴的侧面下陷的孢子囊

Fertile spike
可育孢子囊穗

Whole plant of (*Ophioglossum vulgatum*)
瓶尔小草的整个植株

Clustered sporangia on a sporophyll (*Osmunda regalis*)
孢子叶上的丛生孢子囊（欧紫萁）

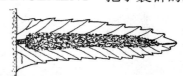

Coensorus along midrib (*Blechnum* sp.)
沿中脉的聚合孢子囊群（乌毛蕨属某植物）

Coensorus on margin (*Pyrrosia* sp.)
边缘的聚合孢子囊群（石韦属某植物）

Elongate on pinna margin (*Pteris* sp.)
羽片边缘长形的（凤尾蕨属某植物）

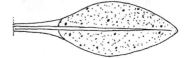

Acrostichoid (*Elaphoglossum* sp.)
卤蕨型的（舌蕨属某植物）

Ferns 6, Fern Allies 1
蕨类 6，拟蕨类 1

Caudex
茎

Tree fern (*Cyathea australis*)
树蕨（桫椤）

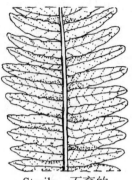

Sterile　不育的　　　　Fertile　可育的

The dimorphic fronds of *Blechnum* sp.
乌毛蕨属某植物的两型叶

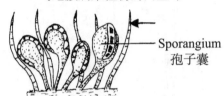

Sporangium
孢子囊

Paraphyses among fern sporangia
蕨类植物孢子囊间的隔丝

TYPES OF STELE
中柱类型

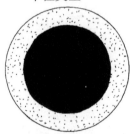

Haplostele
(*Gleichenia* sp.)
单中柱
（里白属某植物）

Actinostele
(*Lycopodium serratum*)
星状中柱
（蛇足石松）

Mixed protostele
(*Lycopodium cernuum*)
混合原生中柱
（铺地蜈蚣草）

Plectostele
(*Lycopodium clavatum*)
编织中柱
（东北石松）

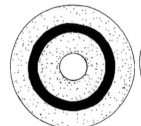
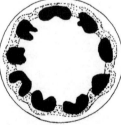
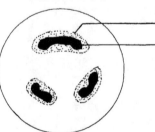

Phloem
韧皮部 } Meristele
Xylem 分体中柱
木质部

Amphiphloic siphonostele
(*Marsilea* sp.)
双韧管状中柱
（苹属某植物）

Ectophloic siphonostele
(*Osmunda regalis*)
外韧管状中柱
（欧紫萁）

Dictyostele
(*Dryopteris* sp.)
网状中柱
（鳞毛蕨属某植物）

Fern Allies 2
拟蕨类 2

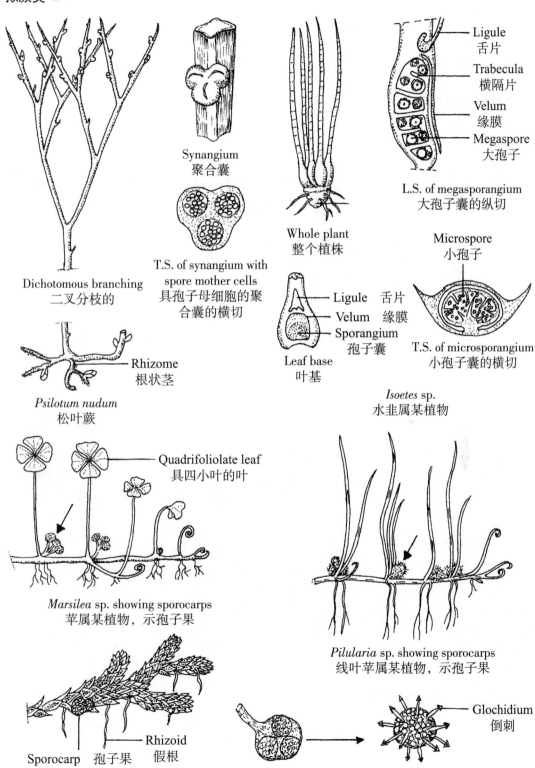

Synangium
聚合囊

T.S. of synangium with
spore mother cells
具孢子母细胞的聚
合囊的横切

Dichotomous branching
二叉分枝的

Whole plant
整个植株

Ligule   舌片
Velum   缘膜
Sporangium
孢子囊

Leaf base
叶基

Ligule
舌片

Trabecula
横隔片

Velum
缘膜

Megaspore
大孢子

L.S. of megasporangium
大孢子囊的纵切

Microspore
小孢子

T.S. of microsporangium
小孢子囊的横切

Rhizome
根状茎

Psilotum nudum
松叶蕨

Isoetes sp.
水韭属某植物

Quadrifoliolate leaf
具四小叶的叶

Marsilea sp. showing sporocarps
苹属某植物，示孢子果

Pilularia sp. showing sporocarps
线叶苹属某植物，示孢子果

Rhizoid
假根

Sporocarp   孢子果

Azolla sp.
满江红属某植物

Microsporangium with massulae
具花粉小块的小孢子囊

Glochidium
倒刺

Massula
花粉小块

Fern Allies 3 *Lycopodium clavatum*
拟蕨类 3　东北石松

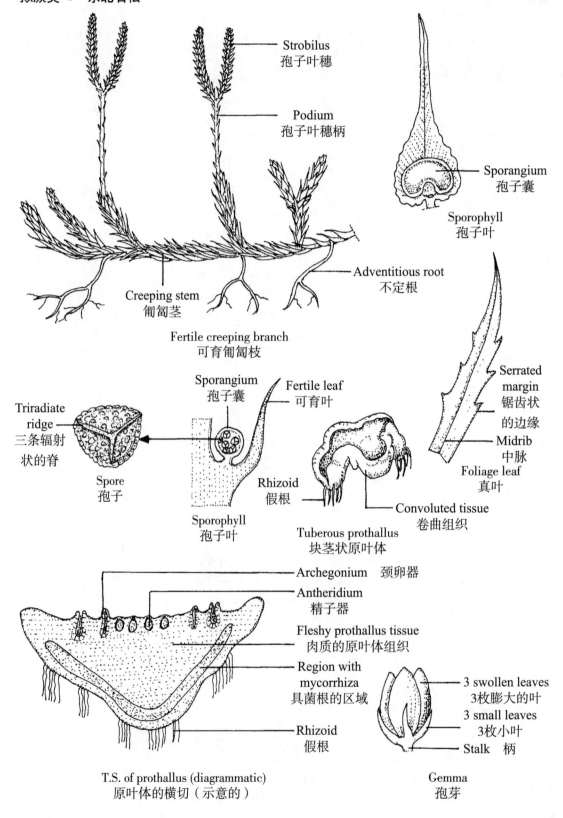

Strobilus
孢子叶穗

Podium
孢子叶穗柄

Creeping stem
匍匐茎

Adventitious root
不定根

Fertile creeping branch
可育匍匐枝

Sporangium
孢子囊

Sporophyll
孢子叶

Sporangium
孢子囊

Fertile leaf
可育叶

Serrated
margin
锯齿状
的边缘

Triradiate
ridge
三条辐射
状的脊

Spore
孢子

Rhizoid
假根

Midrib
中脉

Foliage leaf
真叶

Sporophyll
孢子叶

Convoluted tissue
卷曲组织

Tuberous prothallus
块茎状原叶体

Archegonium　颈卵器

Antheridium
精子器

Fleshy prothallus tissue
肉质的原叶体组织

Region with
mycorrhiza
具菌根的区域

Rhizoid
假根

3 swollen leaves
3枚膨大的叶

3 small leaves
3枚小叶

Stalk　柄

T.S. of prothallus (diagrammatic)
原叶体的横切（示意的）

Gemma
孢芽

Fern Allies 4 *Equisetum*
拟蕨类 4 木贼属

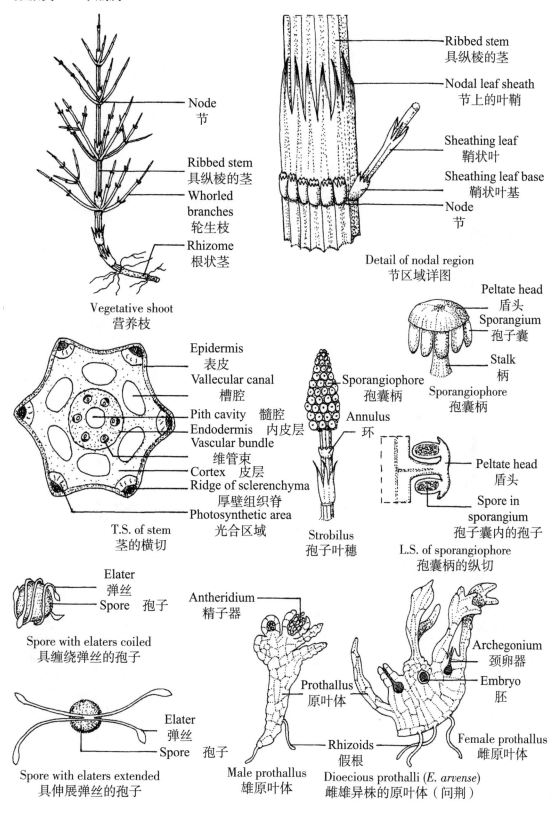

Node
节

Ribbed stem
具纵棱的茎

Whorled
branches
轮生枝

Rhizome
根状茎

Vegetative shoot
营养枝

Ribbed stem
具纵棱的茎

Nodal leaf sheath
节上的叶鞘

Sheathing leaf
鞘状叶

Sheathing leaf base
鞘状叶基

Node
节

Detail of nodal region
节区域详图

Peltate head
盾头

Sporangium
孢子囊

Stalk
柄

Sporangiophore
孢囊柄

Epidermis
表皮

Vallecular canal
槽腔

Pith cavity   髓腔
Endodermis   内皮层
Vascular bundle
维管束
Cortex   皮层
Ridge of sclerenchyma
厚壁组织脊
Photosynthetic area
光合区域

T.S. of stem
茎的横切

Sporangiophore
孢囊柄

Annulus
环

Strobilus
孢子叶穗

Peltate head
盾头

Spore in
sporangium
孢子囊内的孢子

L.S. of sporangiophore
孢囊柄的纵切

Elater
弹丝

Spore 孢子

Spore with elaters coiled
具缠绕弹丝的孢子

Elater
弹丝

Spore 孢子

Spore with elaters extended
具伸展弹丝的孢子

Antheridium
精子器

Prothallus
原叶体

Rhizoids
假根

Male prothallus
雄原叶体

Archegonium
颈卵器

Embryo
胚

Female prothallus
雌原叶体

Dioecious prothalli (*E. arvense*)
雌雄异株的原叶体（问荆）

Fern Allies 5 *Selaginella*
拟蕨类 5　卷柏属

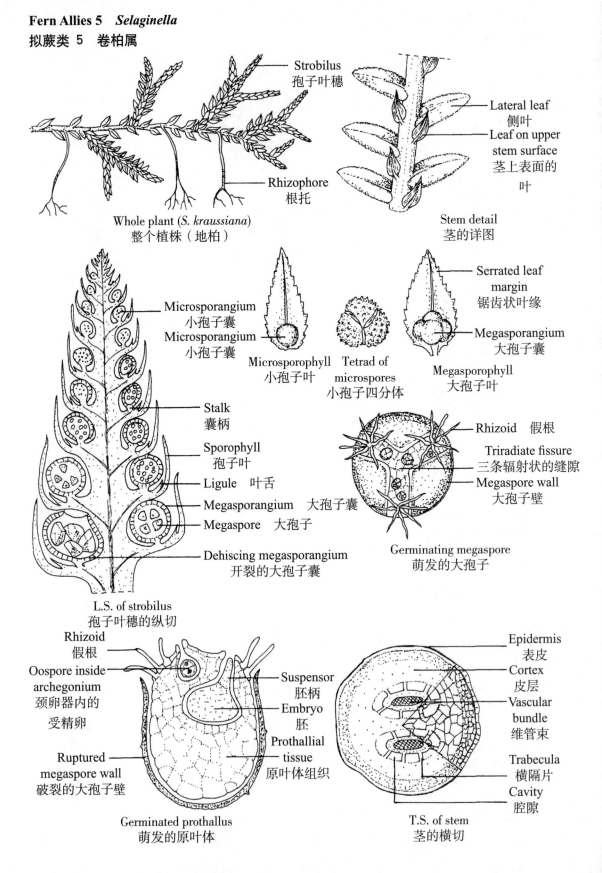

Strobilus
孢子叶穗

Rhizophore
根托

Whole plant (*S. kraussiana*)
整个植株（地柏）

Lateral leaf
侧叶

Leaf on upper stem surface
茎上表面的叶

Stem detail
茎的详图

Microsporangium
小孢子囊
Microsporangium
小孢子囊

Microsporophyll
小孢子叶

Tetrad of microspores
小孢子四分体

Serrated leaf margin
锯齿状叶缘

Megasporangium
大孢子囊

Megasporophyll
大孢子叶

Stalk
囊柄

Sporophyll
孢子叶

Ligule　叶舌

Megasporangium　大孢子囊

Megaspore　大孢子

Dehiscing megasporangium
开裂的大孢子囊

L.S. of strobilus
孢子叶穗的纵切

Rhizoid　假根

Triradiate fissure
三条辐射状的缝隙

Megaspore wall
大孢子壁

Germinating megaspore
萌发的大孢子

Rhizoid
假根

Oospore inside archegonium
颈卵器内的受精卵

Ruptured megaspore wall
破裂的大孢子壁

Germinated prothallus
萌发的原叶体

Suspensor
胚柄

Embryo
胚

Prothallial tissue
原叶体组织

Epidermis
表皮

Cortex
皮层

Vascular bundle
维管束

Trabecula
横隔片

Cavity
腔隙

T.S. of stem
茎的横切